Fundamentals of Guided-Wave Optoelectronic Devices

Optoelectronic guided-wave devices are used in a wide range of optical fiber communication and optoelectronic systems. In such networks, the electrical and the optical characteristics of guided-wave devices, and the interplay between them, have a profound effect on system design and overall performance.

Uniquely, this book combines both the optical and electrical behavior of guided-wave optoelectronic devices so that the interwoven properties, including interconnections to external components, are easily understood. It provides the key concepts and analytical techniques that readers can apply to current and future devices. It also presents the impact of material properties on guided-wave devices, and emphasizes the importance of time-dependent interactions between electrical and optical signals. The properties of the devices are presented and compared in terms of system requirements in applications. This is an ideal reference for graduate students and researchers in electrical engineering and applied physics departments, as well as practitioners in the optoelectronics industry.

William S. C. Chang is an Emeritus Professor of the Department of Electrical and Computer Engineering, University of California, San Diego. After receiving his Ph.D. from Brown University in 1957, he pioneered maser and laser research at Stanford University, and he has been involved in guided-wave research since 1971. He has published over 200 technical papers and books, including *Principles of Lasers and Optics* (Cambridge, 2005) and *RF Photonic Technology in Optical Fiber Links* (Cambridge, 2002).

Fundamentals of Guided-Wave Optoelectronic Devices

WILLIAM S. C. CHANG

University of California, San Diego

CAMBRIDGE
UNIVERSITY PRESS

CAMBRIDGE UNIVERSITY PRESS
Cambridge, New York, Melbourne, Madrid, Cape Town, Singapore, São Paulo, Delhi

Cambridge University Press
The Edinburgh Building, Cambridge CB2 8RU, UK

Published in the United States of America by Cambridge University Press, New York

www.cambridge.org
Information on this title: www.cambridge.org/9780521868235

First published 2010

Printed in the United Kingdom at the University Press, Cambridge

A catalog record for this publication is available from the British Library

ISBN 978-0-521-86823-5 Hardback

Contents

Preface

Optoelectronic guided-wave devices are used in many optical fiber communication and optoelectronic systems. In these systems optical and electrical signals are transmitted, received, multiplexed and converted by means of a variety of procedures. In guided-wave optoelectronic devices, laser radiation propagates in a waveguide and energy can be coupled effectively to and from single mode optical fibers. The properties of materials used to fabricate the waveguides have a profound effect on the phase, amplitude or directional variations of the optical waves used for the generation, modulation, switching, conversion, multiplexing, and detection of optical signals. The small lateral dimensions of the waveguide structures provide for efficient control of their optical properties by means of electrical voltages or currents. On the other hand, optical signals are converted back into electrical signals via detectors. Therefore, the electrical characteristics of these devices are as important as their optical properties. Devices may potentially be monolithically integrated optically on the same chip. This is called photonic integration. Optical components may also be integrated, monolithically, with electronic devices on the same chip. This is called optoelectronic integration. In earlier times, these were called integrated optical devices, as opposed to integrated electronic devices.

The manner in which different material properties affect the electrical characteristics as well as the propagation of optical signals in optoelectronic devices is of great importance. Also of considerable importance is the process of back and forth conversion of the electrical signals and of the optical signals. Furthermore, because the electrical signals must be received or transmitted to external circuits, how the devices are interconnected or driven by other electrical systems is also of great importance. The electrical signals may propagate at microwave frequencies within the optoelectronic devices. Therefore their performances must be analyzed and evaluated in terms of time-dependent interactions of electrical and optical waves.

A large number of books are already available in the technical literature on the optical analysis of waveguides. There are also many books that analyze the specific properties of electrical devices and circuits. This book is intended for use as a graduate level reference or text book. It provides an analysis of guided-wave devices from both the optical and the electrical points of view so that the interwoven optical and electrical properties of the devices, including their optical and electrical interconnections to external components, can be represented clearly. When appropriate, the impact of material properties on guided-wave devices is presented and the importance of time-dependent interactions between electrical and optical signals is emphasized. The book emphasizes fundamental concepts and analytical techniques rather than giving a comprehensive coverage of different devices. The intention of the author is to illustrate these concepts and analytical techniques clearly so that they can be applied to all guided-wave optoelectronic devices, including many that are not covered in this book, or have not been investigated as yet.

Optical waveguides can be divided into planar waveguides (two-dimensional) and channel (three-dimensional) waveguides. The fabrication and analysis of optical waveguides constitute the most basic knowledge needed for understanding and designing guided-wave components. Chapter 1 begins with a discussion of the formation and the modal analysis of planar and channel optical waveguides. The optical analysis presented is similar to those in other books concerned with waveguides. Differently from other guided-wave books, a two-dimensional Green's function approach is presented which could be used to analyze propagation of planar guided waves in general. Also included is a description of the materials technologies employed for fabrication of optical waveguides.

The mathematical analysis of channel waveguide modes is already complicated because of the geometry of the boundaries of waveguides. Yet, in order to understand guided-wave devices, it is necessary to analyze the mutual interactions between optical waves in two or more channel waveguides. Therefore approximation techniques such as perturbation and coupled mode analyses are introduced in Chapter 2. They could be used to analyze the coupled waveguides and the interaction of optical waves with changes in material properties. Examples of waveguide components, such as the grating filter, the directional coupler and the Mach–Zehnder interferometer, are used as examples to illustrate these approximate analytical techniques. Another powerful technique useful for analysis of multiple waveguide components is the super mode analysis. It is introduced next in Chapter 2, after the coupled mode analysis. Additional insight into the properties of guided-wave devices such as the directional coupler, the Y-branch coupler and the interference coupler can be obtained from super mode analysis.

Optical amplification and photo-carrier generation are the basis of lasers and photo-detectors and they are described in many other books. In this book, how changing the material properties affects the propagation and interaction of optical waves, thereby producing modulation, switching, beam scanning, etc. in optoelectronic components is treated in detail. Electro-optical effects such as the linear electro-optic effect, the electro-absorption effect and the electro-refraction effect are discussed in Chapter 3.

In optoelectronic applications, electrical fields are created by time varying electrical voltages applied to electrode structures of the components. Analytical techniques for dealing with the time varying electrical properties of optoelectronic guided-wave structures are reviewed in Chapter 4. These techniques include the analyses of electrical fields produced by time varying voltages, the electro-optical effects produced by the electric fields, and the representation of the parameters of electro-optic devices by lumped circuit elements at lower frequencies and by traveling wave transmission lines at higher frequencies. Discussion in this chapter includes issues related to impedance matching such devices to microwave sources. Note that the frequency response and the electrical behavior of the device, in turn, place additional demands on the design of electrode and waveguide configurations.

Chapters 5 and 6 provide a description of guided-wave devices using planar and channel waveguides. The analyses of these devices utilize all the optical and electrical analytical tools, material properties and electro-optical effects described in Chapters 1 to 4. The optical and electrical performances of such devices are evaluated from the application point of view and the properties of different devices designed for the same application are compared to each other.

In planar waveguides, optical guided waves can propagate in any direction, following the contour of the waveguide layer. Summations of planar guided waves form divergent, converging, diffracted and deflected waves. Therefore, how to harness the refraction, diffraction and reflection of planar guided waves by planar waveguide devices is also the focus of Chapter 5. However, most of the applications will involve channel waveguide devices because of the ease of coupling to optical fibers, the superior electro-optical performance derived from the small lateral dimension of channel waveguides, and the advantage of small electrical capacitance of the device at high electrical frequencies. Devices that perform the same practical functions such as power division, wavelength filtering, resonance filtering, signal time delay, switching, multiplexing, and modulation are described, analyzed, and compared together. Their time-dependent characteristics are derived from combined microwave and optical analyses. Device performances are evaluated in terms of the systems requirements in applications. This is an unusual feature of the book.

Acknowledgement

The author is indebted to Professors H. H. Wieder and Paul K. L. Yu at the University of California, San Diego for reviewing the manuscript. Our mission as a university is to explain the basic principles of guided-wave optoelectronics as best we can, and to continue to improve our explanations. The author welcomes any comments from readers. Comments can be sent directly to wchang@ucsd.edu.

1 The formation and analysis of optical waveguides

1.1 Introduction to optical waveguides

Optical waveguides are made from material structures that have a core region which has a higher index of refraction than the surrounding regions. Guided electromagnetic waves propagate in and around the core. The transverse dimensions of the core are comparable to or smaller than the optical wavelength. Figure 1.1(a) illustrates a typical planar waveguide. Figure 1.1(b) illustrates a typical channel waveguide. For rigorous electromagnetic analysis of such guided-wave structures, Maxwell's vector equations should be used. Many of the theoretical methods used in the analysis of optical guided waves are very similar to those used in microwave analysis. For example, modal analysis is again a powerful mathematical tool for analyzing many devices, applications and systems.

However, there are also important differences between optical and microwave waveguides. In microwaves, we usually have closed waveguides inside metallic boundaries. Metals are considered as perfect conductors at most microwave frequencies. Microwaves propagate within the metallic enclosure. Figure 1.2 illustrates a typical microwave rectangular waveguide. In these closed structures, we have only a discrete set of waveguide modes whose electric fields terminate at the metallic boundary. Microwave radiation in the waveguide may be excited either by an electric field or by a current loop. At optical wavelengths, we avoid the use of metallic boundaries because of their strong absorption of radiation. Ideal optical waveguides, such as those illustrated in Fig. 1.1(a) and (b), are considered to have dielectric boundaries extending to infinity. They are called open waveguides. Optical guided-wave modes are waves trapped in and around the core. They can be excited only by electric fields.

1.1.1 Differences between optical and microwave waveguides

Mathematically, modes represent propagating homogeneous[1] solutions of Maxwell's electromagnetic equations in waveguide structures that have constant cross-section and infinite length. Homogeneous solutions means that these are the propagating electric and magnetic fields that satisfy the differential equations and all the boundary conditions in the absence of any radiation source.[2] There are three important differences between optical and microwave waveguide modes and their utilization.

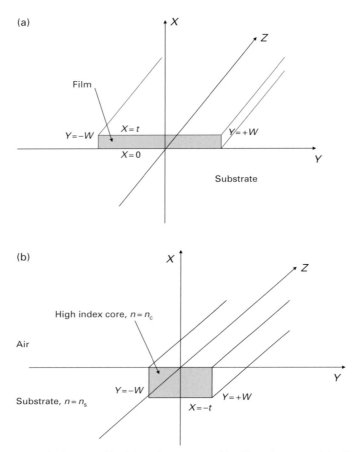

Fig. 1.1.   An optical waveguide. (a) A planar waveguide. The substrate and the film are so wide in the Y direction that W can be approximated by ∞. The substrate thickness is also considered to be ∞ in the $-x$ direction. Guided-wave modes could propagate in any direction in the YZ plane. (b) A channel waveguide. The high index core $(-t \leq x \leq 0, -W \leq y \leq +W)$ is embedded in the substrate. The core is very long in the z direction with $n_c > n_s > 1$. The guided wave propagates in the z direction.

(1) In open dielectric waveguides, the discrete optical modes have an evanescent field outside the core region (the core is often called vaguely the optical waveguide). There may be a significant amount of energy carried in the evanescent tail. The evanescent field may be used to achieve mutual interactions with the fields of other modes of such waveguides or structures. The evanescent field interaction is very important in devices such as the dielectric grating filter, the distributed feedback laser and the directional coupler.

(2) The mathematical analysis is more complex for open than for closed waveguides. In fact, there exists no analytical solution of three-dimensional open channel waveguide modes (except the modes of the round step index fiber) in the closed form. One must use either numerical analysis or approximate solutions in order to find the field distribution of optical channel waveguide modes.

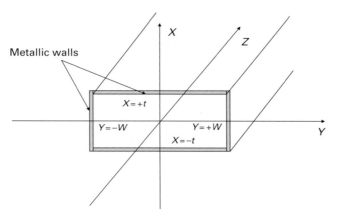

Fig. 1.2. A microwave waveguide. The rectangular waveguide has metallic walls at $y = \pm W$ and at $x = \pm t$. Guided waves propagate along the Z direction in the hollow region, $-t < x < +t$, $-W < y < +W$.

(3) In addition to the set of guided modes that have discrete eigenvalues, there is an infinite set of continuous modes in open waveguides. Only the sum of the discrete and continuous modes constitutes a complete set of orthogonal functions. It means that, rigorously, any arbitrary incident field should be expanded mathematically as a summation of this complete set of modes. At any dielectric discontinuity, the boundary conditions of the continuity of electric and magnetic fields are satisfied by the summation of both the guided-wave modes and the continuous modes on both sides of the boundary. In other words, continuous modes are excited at any discontinuity. Energy in the continuous modes is radiated away from the discontinuity. Thus, continuous modes are called radiation modes.

1.1.2 Diffraction of plane waves in waveguides

The propagation and properties of optical waves in optical waveguides can also be understood from conventional optical analysis of plane wave propagation in multilayered media. A typical optical planar waveguide is illustrated in Fig. 1.3. It has a high index film surrounded by cladding and a substrate; both have a lower index of refraction. The width of the film, the cladding and the substrate, extend to $y = \pm\infty$. The thickness of the substrate and cladding also extends to infinity in the x direction. If we analyze optical plane waves propagating in multilayered media such as that shown in Fig. 1.3, we find that there are three typical cases.

(1) In the first of these, a plane wave is incident obliquely on the film from either $x \ll 0$ or $x \gg t$. Without any loss of generality, let us assume that the plane wave is polarized in the y direction. It propagates in the xz plane in a direction which makes an angle θ_j with respect to the x axis. The angle, θ_j, will be different in different layers, where j designates the layer with index n_j. For example, plane waves in the film with index n_1 will have a functional form, $\exp(\pm jn_1 k \sin\theta_1 z)\exp(\pm jn_1 k \cos\theta_1 x)\exp(j\omega t)$.

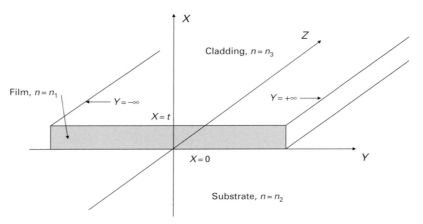

Fig. 1.3.   The index profile in a planar waveguide.

There will be reflected and transmitted waves at the top and bottom boundaries of the film. The continuity of the tangential electric field demands that $n_1 k \sin \theta_1 = n_2 k \sin \theta_2 = n_3 k \sin \theta_3$ at the boundaries. There is a continuous range of real values of θ_j that will satisfy Maxwell's equations and the boundary conditions in all the layers. Plane waves with real values of θ_j represent radiation waves for $x < 0$ and for $x > t$ because they propagate in the x direction. In the language of modal analysis, the multiple reflected and refracted waves constitute the radiation modes with continuous eigenvalues β_{xj} ($\beta_{xj} = n_j k \cos \theta_j$) in the x direction, and β_{xj} is real.

(2) In the second cases, the y-polarized plane waves are trapped in the high index film by total internal reflections from the top and the bottom boundaries of the film at $x = 0$ and $x = t$. In this case the plane waves in the film will still have the functional variation of $\exp(\pm jn_1 k \, \sin \theta_1 z) \exp(\pm jn_1 k \, \cos \theta_1 x) \exp(j\omega t)$ with real values of θ_1. When θ_1 is sufficiently large, total internal reflection occurs at the boundaries. In total internal reflection, "$n_1 k \sin\theta_1$" is larger than $n_j k$ of the surrounding media, and θ_j (for $j \neq 1$) becomes imaginary in order to satisfy the boundary conditions at all values of z. The fields in the cladding and substrate regions, $x < 0$ and $x > t$, decay exponentially away from the boundaries. When the trapped waves in the high index film are bounced back and forth between the two boundaries, they will cancel each other because of the difference in phase and yield zero total field except at specific values of θ_1 at which the round trip phase shift of the reflections is a multiple of 2π. In other words, trapped waves can only have discrete values of real propagation constant, β_{x1} ($\beta_{x1} = n_1 k \cos\theta_1$), in the film in the x direction. It means that plane waves in the substrate and cladding (or air) only have discrete imaginary θ_j values outside the film. As we shall show later, *the non-zero (i.e. the homogeneous solutions of wave equations) waves trapped in the high index film at these specific θ_1 values constitute the various orders of guided waves. Each order of guided wave propagates in the z direction with a phase velocity equal to ω/β_1* ($\beta_1 = n_1 k \sin \theta_1$).

(3) Let us assume that, in the third situation, the index of the substrate n_2 is higher than the index of the cladding n_3, and lower than the index of the film n_1. In this case, there are propagating plane waves in two regions of x: in the substrate and in the high index film region. The value of θ_1 is just large enough so that plane waves are totally internally reflected at the boundary between the film and the top cladding. Only the field in the cladding region now decays exponentially away from the film boundary.

When there are also index variations in the lateral direction (i.e. the y direction) similar observations, like those we discussed in (2), can be made for optical planar guided waves propagating in the lateral direction in the yz plane. Guided-wave modes in a channel waveguide such as the one shown in Fig. 1.1(b) can be analyzed as planar guided-wave modes totally internal reflected at the lateral boundary at $y = \pm W$, see Section 1.2.6. There will be evanescent fields in the y direction at $y > W$ and $y < -W$.

1.1.3   General characteristics of guided waves

In summary, optical waveguides always have a higher index core, surrounded by lower index regions, so that optical guided waves in the core can be considered as waves trapped in the core with evanescent field in the surrounding regions. There are also radiation waves (or cladding waves) that also propagate in the structure. The field distribution and the propagation constant of the guided waves are controlled by the transverse dimensions of the core and the refractive indices of the core and all the surrounding regions. In order to understand more clearly the properties of modes in the optical waveguide, electromagnetic analysis of modes in optical waveguides is presented in the next section.

The most important characteristics of guided-wave modes are the exponential decay of their evanescent tails, the distinct polarization associated with each mode, and the excitation of continuous modes at any defect or dielectric discontinuity that causes diffraction loss of the guided-wave mode. The evanescent tail ensures that there is only minor perturbation of the mode pattern for structure changes several decay lengths away from the surface of the high index layer.

Since propagation loss of the guided-wave modes is caused usually by scattering or absorption, the attenuation rate of the guided mode will be very low as long as there is very little absorption or scattering loss in or near the high index layer. The most common causes of absorption loss are the placement of a metallic electrode nearby, the absorption of the core material, and the use of semiconductor cladding or substrate (or core and cladding) that has absorption. In electro-absorption modulators or switches (discussed in Section 3.2) the absorption of the waveguide is controlled by an electrical signal so that the output optical power is modulated by the electrical signal. Besides absorption, the propagation losses are most commonly caused by volume scattering in the layers or by surface scattering at the dielectric interfaces. Volume scattering is introduced by defects in the material developed during growth or processing deposition. Surface scattering is created usually through roughness incurred in the fabrication processes such as etching and lift-off. Scattering converts the energy in the guided-wave mode into radiation modes.

The exponential decay rate of any guided-wave mode in the media surrounding the core is determined only by the indices of the layers (e.g. either the cladding index at $x > t$ or the substrate index at $x < 0$, in planar waveguides) and the β value of the mode. The $\beta c/\omega$ value is called the effective index, n_{eff}, of the mode. The velocity of light in free space c divided by the effective index is the phase velocity of the guided-wave mode in the z direction. For the same polarization, lower order modes will have a larger effective index (i.e. larger β) and faster exponential decay outside the core. For the same defects or interface roughness, modes that have a smaller effective index will be scattered more strongly into radiation modes. Therefore, higher order modes usually have larger attenuation. Any mode that has an effective index very close to the refractive index of the substrate or cladding will have large scattering loss. It is called a weakly propagating mode.

On the other hand, the evanescent tail also enables us to affect the propagation of the guided-wave mode by placing perturbations adjacent to the core of the high index layer. For example, in the next chapter, we will discuss the directional coupler formed by two waveguides placed adjacent to each other or by a grating filter fabricated on top of a waveguide.

1.2   Electromagnetic analysis of modes in optical waveguides

In order to understand clearly the electromagnetic properties of guided waves, modal analysis of an optical waveguide is presented in this section. The rigorous mathematical analysis of simple planar waveguides such as those shown in Fig. 1.1(a) will be presented first. In principle, modes of planar waveguides (or a summation of planar guided-wave modes) may propagate in any direction in the plane of the waveguide (i.e. the yz plane). However, for simplicity and without any loss of generality, the mathematical solution of the modes of the planar waveguide will be presented first just for modes propagating in the z direction. How these modes of planar waveguide (or combination of modes) propagate in any arbitrary direction in the yz plane will be discussed in terms of these z-propagating modes.

The geometry of channel waveguides is usually too complex for us to find mathematically the solutions of the Maxwell's equations in closed form. Numerical simulation programs such as *Rsoft BeamProp©* are used. The exception is the solution of the circular symmetric modes in step-index round fibers. The modes of optical fibers have been discussed in many books [1]. They will not be repeated here. We will discuss in Section 1.2.6 an approximate analysis, called the effective index analysis, of the modes of open rectangular channel waveguides such as those shown in Fig. 1.1(b). Results obtained from the effective index analysis are accurate only for well-guided modes, i.e. modes with a short evanescent tail. Nevertheless, the effective index analysis enables us to understand the basic properties of all channel guided-wave modes.

It will be clear later from the discussions of planar and channel waveguide modes that the fields of most guided-wave modes can be approximated just by the dominant component of the mode perpendicular to the direction of propagation. In other words, instead of solving Maxwell's vector equations, modes of arbitrary cross-section of the core may be calculated approximately by a scalar equation in terms of just the dominant field. Such a quasi-scalar approximation of the Maxwell's equations will be presented after the discussion of planar and channel waveguide modes.

1.2.1 The asymmetric planar waveguide

A typical uniform dielectric thin film planar waveguide has been shown in Fig. 1.3, where the film, the cladding and the substrate are all uniform and infinitely wide in the y and the z directions. The film typically has a thickness of the order of a wavelength or less, supported by a substrate and a cladding many wavelengths (or infinitely) thick. The refractive index of the film (i.e. the waveguide core), n_1, is higher than the indices of the surrounding layers.

Since the structure is identical in any direction in the yz plane, we will temporarily choose the $+z$ axis as the direction of propagation in our mathematical analysis. For planar modes, we further assume $\partial/\partial y \equiv 0$. This assumption is similar to the assumption made for plane waves in a homogeneous medium in many textbooks. This assumption on the y variation applies in Sections 1.2.2, 1.2.3 and 1.2.4.

1.2.2 TE and TM modes in planar waveguides

The variation of the refractive index in the transverse direction is independent of z in Fig. 1.3. From discussions of electromagnetic theory in classical electrical engineering textbooks, we know that modes for structures that have constant transverse cross-section in the direction of propagation can be divided into TE (transverse electric) and TM (transverse magnetic) types. Note that TE means that there is no electric field component in the direction of propagation, TM means that there is no magnetic field component in the direction of propagation.

For planar waveguides, if we substitute $\partial/\partial y = 0$ into $\nabla \times \underline{E}$ and $\nabla \times \underline{H}$ in Maxwell's equations, we obtain two separate groups of equations:

$$\frac{\partial E_y}{\partial z} = \mu \partial H_x/\partial t, \quad \frac{\partial E_y}{\partial x} = -\mu \partial H_z/\partial t, \quad \frac{\partial H_z}{\partial x} - \frac{\partial H_x}{\partial z} = -\varepsilon \partial E_y/\partial t,$$

and

$$\frac{\partial H_y}{\partial z} = -\varepsilon \partial E_x/\partial t, \quad \frac{\partial H_y}{\partial x} = \varepsilon \partial E_z/\partial t, \quad \frac{\partial E_z}{\partial x} - \frac{\partial E_x}{\partial z} = \mu \partial H_y/\partial t. \tag{1.1}$$

Clearly, E_y, H_x, and H_z are related only to each other, and H_y, E_x, and E_z are related only to each other. Since the direction of propagation is z, the solutions of the first group of equations are the TE modes. The solutions of the second group of equations are the TM modes. In other words, all planar waveguide modes can be divided into TE and TM types.

Since ε is only a function of x, the z variation of the fields must be the same in all layers. This is a consequence of the requirement for continuity of E_y or H_y for all z. Let us also assume that the time variation of the field is $e^{j\omega t}$. Then, for propagating waves in the $+z$ direction, we will have an $\exp(-j\beta z)$ variation, while the waves in the $-z$ direction will have an $\exp(j\beta z)$ variation. The TE wave equations for planar E_y in Eq. (1.1) can now be written as a product of a function in y and a function in z, i.e. $E_y(x,z) = E_y(x)E_y(z)$

$$\left[\frac{\partial^2}{\partial x^2} + \left(\omega^2\mu\varepsilon(x) - \beta^2\right)\right]E_y(x)E_y(z) = 0, \tag{1.2a}$$

$$\left[\frac{\partial^2}{\partial z^2} + \beta^2\right]E_y(z) = 0, \tag{1.2b}$$

or

$$\left[\frac{\partial^2}{\partial x^2} + \left(\omega^2\mu\varepsilon(x) - \beta^2\right)\right]E_y(x) = 0. \tag{1.2c}$$

Similar equations exist for TM modes.

1.2.3   TE modes of planar waveguides

The planar TE modes (i.e. modes with $\partial/\partial y = 0$) in the planar waveguides are eigen solutions of the equation,

$$\left[\frac{\partial^2}{\partial x^2} + \frac{\partial^2}{\partial z^2} + \omega^2\mu\varepsilon(x)\right]E_y(x,z) = 0$$

$$\varepsilon(x) = n_3^2\varepsilon_0 \quad x \geq t$$
$$= n_1^2\varepsilon_0 \quad t > x > 0$$
$$= n_2^2\varepsilon_0 \quad 0 \geq x$$

$$H_x = -\frac{j}{\omega\mu}\frac{\partial E_y}{\partial z}, \quad H_z = \frac{j}{\omega\mu}\frac{\partial E_y}{\partial x}. \tag{1.3}$$

Here, ε_0 is the free space electric permittivity. All layers have the same magnetic permeability μ, and the time variation is $\exp(j\omega t)$. Note that when E_y is known, H_x and H_y can be calculated directly from E_y. The boundary conditions are the continuity of the tangential electric and magnetic fields at $x = 0$ and at $x = t$. As we shall see in the following subsections, the TE modes can be further classified into three sub-groups. One group, the guided waves, is characterized as plane waves trapped inside the film, and the other two groups are two different kinds of combination of radiating plane waves known as substrate modes and air modes. Mathematically, all the TE modes form a complete set of eigenfunctions, meaning that any arbitrary electric field polarized in the y direction with $\partial/\partial y = 0$ can be expanded as a summation of TE modes.

1.2.3.1   TE planar guided-wave modes
Mathematically, Eq. (1.2) and (1.3) suggest that the solution of $E_y(x)$ is either a sinusoidal or an exponential function, and the solution of $E_y(z)$ is $e^{\pm j\beta z}$. Guided by the discussion in

Section 1.1, we look for solutions of $E_y(x)$ with sinusoidal variations for $t > x > 0$ and with decaying exponential variations for $x > t$ and $x < 0$. Since we have chosen the time variation as $e^{+j\omega t}$, the $\exp(-j\beta z)$ variation of $E_y(z)$ represents a forward propagating wave in the $+z$ direction. In short, we will assume the following functional form for $E_y(x,z)$:

$$
\begin{aligned}
E_m(x, z) &= A_m \ \sin(h_m t + \phi_m) \exp[-p_m(x - t)] \exp(-j\beta_m z) & x \geq t \\
E_m(x, z) &= A_m \ \sin(h_m x + \phi_m) \exp(-j\beta_m z) & t > x > 0 \\
E_m(x, z) &= A_m \ \sin \phi_m \exp[q_m x] \exp(-j\beta_m z), & 0 \geq x
\end{aligned}
$$

where in order to satisfy Eq. (1.2a, b and c)

$$
\begin{aligned}
(\beta_m/k)^2 - (p_m/k)^2 &= n_3^2 \\
(\beta_m/k)^2 + (h_m/k)^2 &= n_1^2 \\
(\beta_m/k)^2 - (q_m/k)^2 &= n_2^2.
\end{aligned}
\tag{1.4}
$$

The subscript m stands for the mth order solution of Eq. (1.3). Equation (1.3) is clearly satisfied by E_m in all the individual regions. We have also chosen this functional form so that the continuity of E_y is automatically satisfied at $x = 0$ and $x = t$. In order to satisfy the magnetic boundary conditions[3] at $x = 0$ and $x = t$, h_m, q_m, and p_m must be the mth set of the roots of the transcendental equations which are also called the characteristic equations,

$$
\tan[(h_m/k)kt + \phi_m] = -h_m/p_m \quad \text{and} \quad \tan \phi_m = h_m/q_m.
\tag{1.5}
$$

For a given normalized thickness kt, there are only a finite number of roots of the characteristic equations yielding a discrete set of real values for h, p, and q. For this reason, the guided-wave modes are also called the discrete modes. They are labeled by the integer subscript m ($m = 0, 1, 2,\ldots$). The lowest order mode with $m = 0$ has the largest β value, $\beta_0 > \beta_1 > \beta_2 > \beta_3 \ldots$ and $h_0 < h_1 < h_2 \ldots$. Moreover, one can show that the number of times in which $\sin(h_m x + \phi_m)$ is zero is m. The H_x and H_z fields can be calculated from E_y according to Eq. (1.1). Since $\beta_m >> h_m$, H_x is the dominant magnetic field for TE modes. The mth TE mode propagating in the $-z$ direction will have $e^{j\beta z}$ variation for $E_y(z)$, with the same xy field variation given in Eq. (1.4).

The exponential decay rate of any guided-wave mode is determined only by the index of the surrounding layer (either at $x > t$ or at $x < 0$) and the β/k value of the mode. The β/k value is called the effective index, n_{eff}, of the mode. The velocity of light in free space divided by effective index n_{eff} is the phase velocity of the guided-wave mode. For the same polarization, lower order modes will have larger effective index and faster exponential decay. For the same $\Delta\varepsilon$ of defects or interface roughness, modes that have a smaller effective index will be scattered more strongly into radiation modes, i.e. substrate and air modes. Therefore, higher order modes usually have larger attenuation.

1.2.3.2 TE planar guided-wave mode in a symmetrical waveguide

In order to visualize why there should be only a finite number of modes, let us consider the example of a symmetrical waveguide. In that case, $n_2 = n_3 = n$ and $p_m = q_m$. The quadratic equations for h_m and β_m and the transcendental equation now become

$$\left(\frac{h_m}{k}\right)^2 + \left(\frac{p_m}{k}\right)^2 = n_1^2 - n^2, \tag{1.6}$$

and

$$\tan\left[\left(\frac{h_m}{k}\right)kt\right] = \frac{-2\frac{h_m}{p_m}}{1 - \frac{h_m^2}{p_m^2}}. \tag{1.7}$$

Since,

$$\tan\left[2\left(\frac{h_m}{k}\right)\frac{kt}{2}\right] = \frac{2\,\tan\left[\left(\frac{h_m}{k}\right)\frac{kt}{2}\right]}{1 - \tan^2\left[\left(\frac{h_m}{k}\right)\frac{kt}{2}\right]},$$

Eq. (1.7) can be reduced to two equations,

$$\tan\left[\left(\frac{h_m}{k}\right)\frac{kt}{2}\right] = \frac{p_m/k}{h_m/k}, \quad \text{hence} \quad \frac{h_m}{k}\tan\left[\left(\frac{h_m}{k}\right)\frac{kt}{2}\right] = \frac{p_m}{k}, \tag{1.8a}$$

or

$$\tan\left[\left(\frac{h_m}{k}\right)\frac{kt}{2}\right] = -\frac{h_m/k}{p_m/k}, \quad \text{hence} \quad -\frac{h_m}{k}\cot\left[\left(\frac{h_m}{k}\right)\frac{kt}{2}\right] = \frac{p_m}{k}. \tag{1.8b}$$

In the coordinate system of p_m/k and h_m/k, the solutions of Eq. (1.6) and (1.7) are given by the intersections of the two curves representing the quadratic equation, $(h_m/k)^2 + (p_m/k)^2 = n_1^2 - n^2$, and one of the two equivalent tangent equations, (1.8a) or (1.8b). To summarize, there are two sets of equations. The solutions for the first tangent equation (1.8a) and the quadratic equation (1.6) are known as the even modes because they lead to field distributions close to a cosine variation in the film. They are symmetric with respect to $x = t/2$. The solutions from the second tangent equation (1.8b) and the quadratic equation (1.6) are called the odd modes because the fields in the film have distributions close to sine variations. They are anti-symmetric with respect to $x = t/2$.

Let us examine the even modes in detail. If we plot the quadratic equation of h_m/k and p_m/k, it is a circle with a radius $(n_1^2 - n^2)^{1/2}$. The curve describing the first tangent equation will be obtained from those values of h_m/k and p_m/k whenever the left hand side (LHS) is equal to the right hand side (RHS) of the tangent equation. The RHS is just p_m/k. The LHS has a tangent which is a multi-valued function. It starts from 0 whenever $(h_m/k)kt/2$ is 0, π, or $m\pi$. It approaches + or − infinity when $(h_m/k)kt/2$ approaches $+\pi/2$ or $-\pi/2$, or $(m+\pi/2)$ or $(m-\pi/2)$ where m is an integer. The curves representing these two equations are illustrated in Fig. 1.4. Clearly there is always a solution as long as $n_1 > n$, i.e. there is an intersection of the two curves, no matter how large (or how small) is the circle (i.e. the n_1 value). This is the fundamental mode, labeled by $m = 0$. However, whether there will be an $m \geq 1$ solution depends on whether the radius is larger than $2\pi/kt$. There will be $m = j$ solutions when the radius is larger than $2j\pi/kt$. Notice that $h_0 < h_1 < h_2 \ldots$ and $\beta_0 > \beta_1 > \beta_2 > \ldots$. When the radius of the circle is just equal to $2j\pi/kt$, the value for p/k is 0. This is the cut-off point for the jth $(j > 1)$ mode.

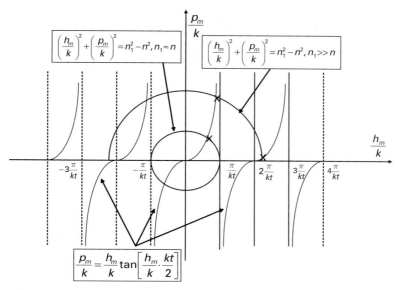

$$\left(\frac{h_m}{k}\right)^2 + \left(\frac{p_m}{k}\right)^2 = n_1^2 - n^2, n_1 \approx n$$

$$\left(\frac{h_m}{k}\right)^2 + \left(\frac{p_m}{k}\right)^2 = n_1^2 - n^2, n_1 \gg n$$

$$\frac{p_m}{k} = \frac{h_m}{k} \tan\left[\frac{h_m}{k} \cdot \frac{kt}{2}\right]$$

Fig. 1.4. The Graphical Solution of h_m and p_m for even TE guided-wave modes in a symmetrical planar waveguide. Taken from ref. 4, Cambridge University Press.

1.2.3.3 The cut-off condition of TE planar guided-wave modes

There are conditions imposed on the refractive indices without which there is no guided-wave mode solution for asymmetric waveguides. The first condition is:

$$n_1 > n_2 \text{ and } n_3.$$

Without any loss of generality, let $n_1 > n_2 \geq n_3$. In addition, there is a minimum thickness t_m, called the cut-off thickness, which will permit the mth solution to Eq. (1.3) to exist. However, differently from the symmetric waveguide for which there is always an $m = 0$ even mode, there is a cut-off condition for even the $m = 0$ mode in asymmetric waveguides. At the cut-off of the mth mode, $q_m = 0$, $\beta_m/k = n_2$, $p_m/k = (n_2^2 - n_3^2)^{1/2}$, $\phi_m = \pm(m+1/2)\pi$, and $h_m = k(n_1^2 - n_2^2)^{1/2}$. Thus the cut-off thickness can be calculated from Eq. (1.5) to be:

$$kt_m = \left\{ \left(m + \frac{1}{2}\right)\pi - \tan^{-1}\left[(n_1^2 - n_2^2)/(n_2^2 - n_3^2)\right]^{1/2} \right\} (n_1^2 - n_2^2)^{-1/2}. \tag{1.9}$$

The thicker the film, the larger the number of guided-wave modes the film can support. For all guided-wave modes above cut-off, $n_1 \geq |\beta_m/k| > n_2$. In most applications, t and the indices of the layers are controlled so that there is only one guided-wave mode in the waveguide.

1.2.3.4 Properties of TE planar guided-wave modes

Figure 1.5 shows the effective index, β_m/k, of TE_m planar guided-wave modes in epitaxially grown waveguides on InP substrates as a function of the waveguide thickness t where $n_2 = 3.10$, $n_1 = n_2 + \Delta n$ and $n_3 = 1$. The abscissa is kt in units of π. The Δn, i.e. $n_1 - n_2$, depends

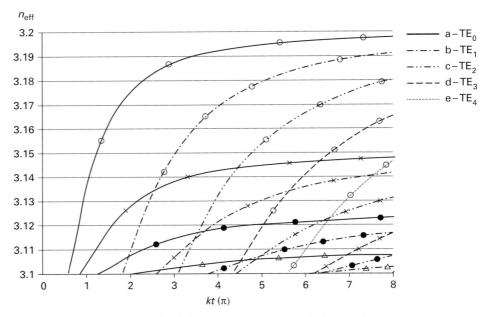

n_{eff}

Fig. 1.5. n_{eff} values of TE_m modes in epitaxially grown waveguides on InP substrates.

on the alloy composition of the epitaxially grown layer. Curves with circles, $\circ$, are for $\Delta n = 0.10$; curves with crosses, $\times$, are for $\Delta n = 0.05$; curves with solid dots, $\bullet$, for $\Delta n = 0.025$; and curves with triangles, $\triangle$, for $\Delta n = 0.01$. The a curves are for TE_0 modes, b curves for TE_1, c curves for TE_2, d curves for TE_3, and e curves for TE_4. At cut-off, all modes have $n_{\text{eff}} = n_2$. The n_{eff} values of the higher modes are always smaller than the n_{eff} values of lower order modes. For a given thickness t, there are more modes for waveguides that have a larger Δn. For $kt < 1.8\pi$, the waveguide has only the TE_0 mode for $\Delta n = 0.1, 0.05$ and 0.025. Notice that we have only real eigenvalues for β, h, p and q. Since β is real, these modes propagate in the z direction without attenuation. The fields of these modes are evanescent in the air and in the substrate. This is the most important characteristic of guided waves.

When there is scattering or absorption loss it usually does not affect the mode pattern significantly. It will cause attenuation as the mode propagates. Figure 1.5 demonstrates clearly that, at a given thickness t, the higher order modes have lower β/k values. Thus the evanescent decay of the higher order modes will be slower in the n_2 and n_3 layers. When there is scattering or absorption loss in the substrate, the slower the evanescent decay, the larger is the attenuation rate. For this reason, higher order guided-wave modes often have a higher attenuation rate. Low scattering and absorption loss in all layers and at all interfaces is a prerequisite for obtaining a low loss waveguide.

Physically, as we have discussed in Section 1.1, the electric field of the mth TE guided-wave mode inside the film is just a plane wave in the n_1 layer (with the electric field polarized in the y direction), totally internally reflected back and forth from the two boundaries at $x = 0$ and $x = t$. Its propagation direction in the xz plane makes an angle θ_m with respect to the x axis:

$$\beta_m = n_1 k \sin \theta_m, \qquad h_m = n_1 k \cos \theta_m. \qquad (1.10)$$

When β_m and h_m are given by the mth solution of Eq. (1.5), the total round trip phase shift of such a plane wave after reflection from both the air and the substrate boundary is $2m\pi$. Since θ_m is a very small angle, the magnetic field of TE modes is polarized predominantly in the x direction with a small component in the z direction. In some technical papers and books, instead of solving Maxwell's equations directly as we did in Eq. (1.4) and (1.5), the guided-wave modes are defined by requiring the round trip phase shift of a totally internally reflected plane wave to be $2m\pi$. This is the condition that the total field for all the plane waves reflected back and forth is non-zero.[4]

The lossless TE planar guided-wave modes are orthogonal to each other and to any other TE or TM modes of the same waveguide [2].[5] It is customary to normalize the constant A_m in Eq. (1.4) so that a unit amount of power (1 W) per unit length in the y direction is carried out by a normalized mode. Thus,

$$\frac{1}{2} \operatorname{Re} \left[\int_{-\infty}^{+\infty} E_{yn} H_{xm}^* \mathrm{d}x \right] = (\beta_m/2\omega\mu) \int E_n E_m^* \mathrm{d}x = \delta_{nm}, \qquad (1.11)$$

where the asterisk superscript of a function means the complex conjugate of that function. From this condition, we obtain

$$A_m^2 = \frac{4\omega\mu}{\beta_m} \left[\frac{1}{p_m} + \frac{1}{q_m} + t \right]^{-1}. \qquad (1.12)$$

1.2.3.5 TE planar substrate modes

In the range, $n_2 > |\beta/k| > n_3$, the electric field has an exponential variation for $x > t$ and sinusoidal variation within the film and in the substrate. According to the discussion given in case (1) of Section 1.1.2, these are substrate modes. For TE substrate modes, we have,

$$E^{(s)}(x,z;\beta) = A^{(s)} \sin(ht + \phi) \exp[-p(x-t)] \exp(-\mathrm{j}\beta z) \qquad x \geq t$$

$$E^{(s)}(x,z;\beta) = A^{(s)} \sin(hx + \phi) \exp(-\mathrm{j}\beta z) \qquad t > x > 0$$

$$E^{(s)}(x,z;\beta) = \left[C^{(s)} \exp(-\mathrm{j}\rho x) + C^{(s)*} \exp(+\mathrm{j}\rho x) \right] \exp(-\mathrm{j}\beta z), \quad 0 \geq x \qquad (1.13)$$

with

$$(h/k)^2 + (\beta/k)^2 = n_1^2$$
$$(\beta/k)^2 - (p/k)^2 = n_3^2$$
$$(\rho/k)^2 + (\beta/k)^2 = n_2^2, \qquad (1.14)$$

$$\tan[(h/t)kt + \phi] = -h/p, \qquad (1.15)$$

and

$$C^{(s)} = A^{(s)}[\sin\phi + \mathrm{j}(h\cos\phi/\rho)]/2. \qquad (1.16)$$

Note that $C^{(s)}$ and $A^{(s)}$ are normalized so that

$$(\beta/2\omega\mu) \int_{-\infty}^{\infty} E^{(s)}(x,z;\beta)E^{(s)*}(x,z;\beta')\mathrm{d}x \;=\; \delta(\rho - \rho'), \tag{1.17}$$

which requires

$$C^{(s)}C^{(s)*} \;=\; \frac{\omega\mu}{\beta\pi}. \tag{1.18}$$

Unlike guided-wave modes, which have $n_1 > |\beta_m/k| > n_2$ and n_3, β, p, h, ρ, and ϕ of the substrate modes have a continuous range of values which satisfy the above equations within the range $n_2 > |\beta/k| > n_3$. Thus these modes are called continuous modes. The field in the air region still has an evanescent variation. However, the field in the substrate region has the form of two propagating plane waves with propagation constant ρ, one in the $+x$ direction and the second one in the $-x$ direction. Thus they are also called the substrate radiation modes.[6]

In the plane wave description of the substrate modes in Section 1.1, $\beta/n_1 k$, $h/n_1 k$, $\beta/n_2 k$ and $\rho/n_2 k$ are direction cosines of the plane waves with respect to the z axis and the x axis in the film region and in the substrate region respectively. The plane waves in the film are totally internally reflected only at the boundary $x = t$.

1.2.3.6   TE planar air modes

Mathematically, there are always two independent solutions of Maxwell's equations for a given set of propagation constants. By linearly combining the two independent solutions one can always obtain two orthogonal independent modes for each set of propagation constants. These orthogonal modes are called air modes because they propagate in both substrate and cladding with index n_2 and n_3. They are called air modes because the cladding medium with n_3 is often the air.

If the structure is symmetrical, these two orthogonal modes represent odd and even variations with respect to $x = t/2$ inside the film. For asymmetrical structures, such as the one shown in Fig. 1.3, the x variations are more complex. Nevertheless, there are still two modes for each set of propagation constants, these two modes differing from each other by a $\pi/2$ phase shift of the sinusoidal variations in the x direction in the film which has the index n_1. The mathematical expressions for E_y of the air modes are:

$$
\begin{aligned}
E'(x,z;\beta) &= \{D' \exp[-j\sigma(x-t)] + D'^* \exp[+j\sigma(x-t)]\}\exp(-j\beta z) & x \geq t\\
E'(x,z;\beta) &= A' \sin(hx+\phi)\exp(-j\beta z) & t > x > 0\\
E'(x,z;\beta) &= [C' \exp(-j\rho x) + C'^* \exp(+j\rho x)]\exp(-j\beta z) & 0 \geq x,
\end{aligned}
\tag{1.19}
$$

for the first set, and

$$
\begin{aligned}
E''(x,z;\beta) &= \{D'' \exp[-j\sigma(x-t)] + D''^* \exp[+j\sigma(x-t)]\}\exp(-j\beta z) & x \geq t\\
E''(x,z;\beta) &= A'' \sin\left(hx+\phi+\frac{\pi}{2}\right)\exp(-j\beta z) & t > x > 0\\
E''(x,z;\beta) &= [C'' \exp(-j\rho x) + C''^*(+j\rho x)]\exp(-j\beta z), & 0 \geq x
\end{aligned}
\tag{1.20}
$$

for the second set, with

$$
\begin{aligned}
(\beta/k)^2 + (\sigma/k)^2 &= n_3^2 \\
(\beta/k)^2 + (h/k)^2 &= n_1^2 \\
(\beta/k)^2 + (\rho/k)^2 &= n_2^2.
\end{aligned}
\tag{1.21}
$$

Imposing the boundary conditions at $x = 0$ and $x = t$, we obtain:

$$
\begin{aligned}
C' &= A'[\sin\phi + \mathrm{j}(h\cos\phi/\rho)]/2 \\
D' &= A'\left[\sin(ht + \phi) + \mathrm{j}\frac{h}{\sigma}\cos(ht + \phi)\right]/2,
\end{aligned}
\tag{1.22}
$$

note that A'', C'', and D'' are obtained when ϕ is replaced by $\phi + \pi/2$ in the above equation. All modes form an orthogonal normalized set as defined in Eq. (1.11) and (1.17). For both sets of modes, a continuous range of solutions of ρ, σ, β and h exists where $n_2 \geq |\beta/k| \geq n_3$.

As discussed in case (1) of Section 1.1.2, the air modes of Eq. (1.4) can be represented in terms of a plane wave with its accompanying reflected and refracted beams at each boundary, without total internal reflection at either boundary. It is well known that for each set of angles of incidence, reflection, and refraction, there are always two independent plane wave solutions. One is a plane wave incident on the film from the cladding side plus its accompanying reflected and refracted waves, and the other is a plane wave incident from the substrate side plus its accompanying reflected and refracted waves. They all have the same z variation. The air modes in Eq. (1.19) and (1.20) are just linear combinations of these plane waves.

1.2.4   TM modes of planar waveguides

The planar TM modes are eigen solutions of the wave equation (with $\partial/\partial y = 0$ and $\exp(\mathrm{j}\omega t)$ time variation):

$$
\begin{aligned}
\left[\frac{\partial^2}{\partial x^2} + \frac{\partial^2}{\partial z^2} + \omega^2\varepsilon(x)\mu\right] H_y(x, z) &= 0 \\
E_x = \frac{\mathrm{j}}{\omega\varepsilon(x)}\frac{\partial H_y}{\partial z}, \quad E_z &= \frac{-\mathrm{j}}{\omega\varepsilon(x)}\frac{\partial H_y}{\partial x},
\end{aligned}
\tag{1.23}
$$

where, $\varepsilon(x)$ is the same as given in Eq. (1.3). Or, in a manner similar to Eq. (1.3), we can write,

$$
\left[\frac{\partial^2}{\partial x^2} + \left(\omega^2\mu\varepsilon(x) - \beta^2\right)\right] H_y(x, z) = 0.
\tag{1.24}
$$

In lossless waveguides, all TM modes are orthogonal to each other and to TE modes [2].

1.2.4.1   TM planar guided-wave modes

Like the TE modes, the y component of the magnetic field for the nth TM planar guided-wave mode is

$$
\begin{aligned}
H_n(x, z) &= B_n \sin(h_n t + \phi_n) \exp[-p_n(x - t)] \exp(-\mathrm{j}\beta_n z) & x \geq t \\
H_n(x, z) &= B_n \sin(h_n x + \phi_n) \exp(-\mathrm{j}\beta_n z) & t > x > 0 \\
H_n(x, z) &= B_n \sin \phi_n \exp[q_n x] \exp(-\mathrm{j}\beta_n z), & 0 \geq x
\end{aligned} \tag{1.25}
$$

with

$$
\begin{aligned}
(\beta_n/k)^2 - (p_n/k)^2 &= n_3^2 \\
(\beta_n/k)^2 + (h_n/k)^2 &= n_1^2 \\
(\beta_n/k)^2 - (q_n/k)^2 &= n_2^2.
\end{aligned} \tag{1.26}
$$

Continuity of the tangential electric field[7] requires that h_n, q_n, and β_n also satisfy the transcendental equation,

$$
\tan[(h_n/k)kt + \phi_n] = -\frac{n_3^2 h_n}{n_1^2 p_n}, \quad \text{and} \quad \tan \phi_n = \left(\frac{n_2}{n_1}\right)^2 \frac{h_n}{q_n}. \tag{1.27}
$$

The TM_n modes are given by the nth solutions of Eq. (1.26) and (1.27). The magnetic field is in the y direction. Note that, differently from the TE guided-wave modes, TM guided-wave modes have the dominant electric field polarized in the x direction.

1.2.4.2   TM planar guided–wave modes in a symmetrical waveguide

It is instructive to see what happens to the TM modes in a symmetrical waveguide, i.e. $n_2 = n_3 = n$. The solution obtained in this example will also be used directly in the effective index method to find the TE modes in channel waveguides. In this case, $p_n = q_n$. The quadratic equation for h_n and β_n and the transcendental equations now become

$$
\left(\frac{h_n}{k}\right)^2 + \left(\frac{p_n}{k}\right)^2 = n_1^2 - n^2, \quad \text{and} \quad \tan\left[\left(\frac{h_n}{k}\right)kt\right] = -\frac{2\dfrac{n^2 h_n}{n_1^2 p_n}}{1 - \left(\dfrac{n^2 h_n}{n_1^2 p_n}\right)^2}.
$$

As we have seen in the case of TE guided-wave modes in symmetrical waveguide structures, the above tangent equation is equivalent to two equations,

$$
\tan\left[\left(\frac{h_n}{k}\right)\frac{kt}{2}\right] = -\frac{n^2 h_n/k}{n_1^2 p_n/k}, \quad \text{and} \quad \tan\left[\left(\frac{h_n}{k}\right)kt\right] = \frac{n_1^2 p_n/k}{n^2 h_n/k}, \tag{1.28a}
$$

or

$$
-\frac{n^2}{n_1^2}\left(\frac{h_n}{k}\right)\cot\left[\left(\frac{h_n}{k}\right)\frac{kt}{2}\right] = \frac{p_n}{k}, \quad \text{and} \quad \frac{n^2}{n_1^2}\left(\frac{h_n}{k}\right)\tan\left[\left(\frac{h_n}{k}\right)\frac{kt}{2}\right] = \frac{p_n}{k}. \tag{1.28b}
$$

These equations again point to the existence of two orthogonal sets of modes, the modes symmetric and anti-symmetric with respect to $t/2$. The $n = 0$ symmetric TM mode has no cut-off thickness t. These equations are very similar to the equations for the TE modes, except for the ratio, $(n/n_1)^2$, which is always smaller than 1. Therefore, for the same order (i.e. $m = n$), the p_n values of the TM modes are slightly smaller than the p_m values of the TE modes for the same thickness t and indices.

1.2.4.3 The cut-off condition of TM planar guided-wave modes

Again, for a given normalized thickness kt, there is only a finite number of discrete modes, labeled by the subscript n ($n = 0, 1, 2 \ldots$) where $h_0 < h_1 < h_2 < h_3 \ldots$ and $n_1 > \beta_0 > \beta_1 > \beta_2 > \beta_3 \ldots > n_2$. The cut-off thickness for the nth TM mode is given by $q = 0$ and by:

$$kt_n = \left\{ n\pi + \tan^{-1} \left[\frac{n_1^2}{n_3^2} \sqrt{\frac{n_2^2 - n_3^2}{n_1^2 - n_2^2}} \right] \right\} \left(n_1^2 - n_2^2 \right)^{-1/2}. \tag{1.29}$$

Note that the cut-off thickness t_n for TM modes is always larger than the cut-off thickness t_m for TE modes of the same order. *Thus it is possible to design the waveguide with appropriate n_1, n_2, n_3 and t so that only the lowest order TE mode can exist.*

1.2.4.4 Properties of TM planar guided-wave modes

Figure 1.6 shows the effective index n_{eff}, i.e. β_m/k, of TM$_m$ planar guided-wave modes in epitaxially grown waveguides on InP substrates as a function of the waveguide thickness t where $n_2 = 3.1$, $n_1 = n_2 + \Delta n$ and $n_3 = 1$. The abscissa is kt in units of π. The Δn, i.e. $n_1 - n_2$, depends on the alloy composition of the epitaxially grown layer. Curves with circles, $\circ$, are for $\Delta n = 0.10$; curves with crosses, $\times$, are for $\Delta n = 0.05$; curves with solid dots, $\bullet$, for $\Delta n = 0.025$;

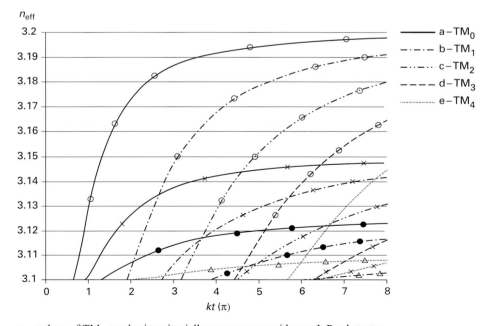

Fig. 1.6. n_{eff} values of TM$_m$ modes in epitaxially grown waveguides on InP substrates.

and with triangles, Δ, for $\Delta n = 0.01$. The a curves are for TM_0 modes, b curves for TM_1, c curves for TM_2, d curves for TM_3, and e curves for TM_4. At cut-off, all modes have $n_{eff} = n_2$. The n_{eff} values of the higher modes are always smaller than the n_{eff} values of lower order modes. For a given thickness t, there are more modes for waveguides that have a larger Δn. For $kt < 0.6\pi$, there is no TM guided wave. Thus a single mode waveguide has only the TE_0 mode. For $kt < 1.9\pi$, the waveguide has only the TM_0 mode for $\Delta n = 0.1$, 0.05 and 0.025. Because of the dependence on $(n_2/n_1)^2$ and $(n_3/n_1)^2$, which are always smaller than 1, the β/k of the TM modes are usually slightly smaller than the corresponding TE modes. *The most important difference between TM and TE modes is, of course, the polarization of the optical electric field.* Often, metallic electrodes are fabricated on top of the n_3 layer intended for applying a DC or RF electric field. The electric field is polarized predominantly in the y direction for TE modes and in the x direction for TM modes. The difference in the polarization of the optical electric field may make a difference to the attenuation of the guided-wave mode in the z direction caused by the metal electrode. For example, when there is metallic absorption, the TM modes have higher attenuation. For other purposes, such as the coupling of an incident radiation field into a planar waveguide, the coupling efficiency depends critically on the matching of the polarization of the incident field with the polarization of the guided-wave mode.

Similarly to TE guided-wave modes, TM planar guided-wave modes inside the film with index n_1 can also be described by a plane wave that has a magnetic field polarized in the y direction. The electric field of the plane wave is predominantly polarized in the x direction with a small component in the z direction. It is totally internally reflected back and forth between the two boundaries, in a propagation direction in the xz plane making an angle θ_n with respect to the x axis. The TM guided-wave modes can be found by requiring the round trip phase shift to be $2n\pi$.

Like the TE modes, the exponential decay rate of any guided-wave mode is determined only by the index of the layer (either at $x > t$ or at $x < 0$) and the β/k value of the mode. The velocity of light in free space c divided by n_{eff} is the phase velocity of the guided-wave mode. For the same polarization, lower order modes will have larger effective index and faster exponential decay. For the same $\Delta\varepsilon$ of defects or interface roughness, modes that have a smaller effective index will be scattered more strongly into radiation modes, i.e. substrate and air modes. Therefore, higher order modes usually have larger attenuation.

When TM guided-wave modes are normalized,

$$\frac{1}{2}\text{Re}\left[\int_{-\infty}^{+\infty} H_{yn}E_{xm}^* \, dx\right] = \frac{\beta_n}{2\omega}\int_{-\infty}^{+\infty} H_n H_m^* \frac{1}{\varepsilon(x)} \, dx = \delta_{nm}, \tag{1.30}$$

and

$$B_n^2 = \frac{4\omega\varepsilon_o}{\beta_n} \cdot \left[\frac{n_1^2}{n_3^2 p_n} \cdot \frac{p_n^2 + h_n^2}{h_n^2 + \left(\frac{n_1^2}{n_3^2}\right)^2 p_n^2} + \frac{n_1^2}{n_2^2 q_n} \cdot \frac{q_n^2 + h_n^2}{h_n^2 + \left(\frac{n_1^2}{n_3^2}\right)^2 q_n^2} + t\right]^{-1}. \tag{1.31}$$

1.2.4.5 TM planar substrate modes

For the substrate TM modes, the y component of the magnetic field is

$$
\begin{aligned}
H^{(s)}(x, z; \beta) &= B^{(s)} \sin(ht + \phi) \exp[-p(x - t)] \exp(-\mathrm{j}\beta z) & x \geq t \\
H^{(s)}(x, z; \beta) &= B^{(s)} \sin(hx + \phi) \exp(-\mathrm{j}\beta z) & t > x > 0 \\
H^{(s)}(x, z; \beta) &= \left[D^{(s)} \exp(-\mathrm{j}\rho x) + D^{(s)*} \exp(+\mathrm{j}\rho x) \right] \exp(-\mathrm{j}\beta z), & 0 \geq x
\end{aligned} \tag{1.32a}
$$

$$
D^{(s)} = \left(\frac{B^{(s)}}{2} \right) \left[\sin\phi + \mathrm{j}\left(\frac{n_2^2 h \cos\phi}{n_1^2 \rho} \right) \right], \tag{1.32b}
$$

$$
\tan[(h/k)kt + \phi] = -\frac{n_3^2 h}{n_1^2 p}. \tag{1.32c}
$$

Values of D and B are obtained from the orthogonalization and normalization condition,

$$
\frac{\beta}{2\omega} \int_{-\infty}^{+\infty} H^{(s)}(\beta) H^{(s)*}(\beta') / \varepsilon(x) \mathrm{d}x = \delta(\rho - \rho'). \tag{1.33}
$$

We obtain

$$
D^{(s)} D^{(s)*} = \frac{\omega \varepsilon_o n_2^2}{\beta \pi}, \tag{1.34}
$$

noting that β, p, h, ρ and ϕ have a continuous range of solutions within the range $n_2 > |\beta/k| > n_3$.

1.2.4.6 TM planar air modes

There are again two orthogonal TM air modes for each set of propagation constants.

For the first set of modes,

$$
\begin{aligned}
H'(x, z; \beta) &= \{ E' \exp[-\mathrm{j}\sigma(x - t)] + E'^* [\mathrm{j}\sigma(x - t)] \} \exp(-\mathrm{j}\beta z) & x \geq t \\
H'(x, z; \beta) &= B' \sin(hx + \phi) \exp(-\mathrm{j}\beta z) & t > x > 0 \\
H'(x, z; \beta) &= [F' \exp(-\mathrm{j}\rho x) + F'^* \exp(\mathrm{j}\rho x)] \exp(-\mathrm{j}\beta z), & 0 \geq x
\end{aligned} \tag{1.35}
$$

and, for the second set of modes

$$
\begin{aligned}
H''(x, z; \beta) &= \{ E'' \exp[-\mathrm{j}\sigma(x - t)] + E''^* (\mathrm{j}\sigma(x - t)) \} \exp(-\mathrm{j}\beta z) & x \geq t \\
H''(x, z; \beta) &= B'' \sin\left(hx + \phi + \frac{\pi}{2} \right) \exp(-\mathrm{j}\beta z) & t > x > 0 \\
H''(x, z; \beta) &= [F'' \exp(-\mathrm{j}\rho x) + F''^* (\mathrm{j}\rho x)] \exp(-\mathrm{j}\beta z). & 0 \geq x
\end{aligned} \tag{1.36}
$$

For both sets of orthogonal modes, a continuous range of solutions of ρ, σ, β and h, exist where $n_3 \geq |\beta/k| \geq 0$. For the first set of modes, the continuity of the electric and magnetic fields at $x = 0$ and $x = t$ requires:

$$E' = \frac{1}{2}B' \left\{ \sin(ht+\phi) + j\frac{h\,n_3^2\cos(ht+\phi)}{\sigma\,n_1^2} \right\}$$

$$F' = \frac{1}{2}B' \left\{ \sin\phi + j\frac{h n_2^2\cos\phi}{\rho\,n_1^2} \right\}. \tag{1.37}$$

For the second set of modes, ϕ is replaced by $\phi + \pi/2$ in Eq. (1.37).

1.2.5   Generalized guided-wave modes in planar waveguides

In Sections 1.2.3 and 1.2.4, we have presented the analysis of planar modes when they propagate in the direction of the z axis. In reality, planar modes for a waveguide structure as shown in Fig. 1.3 can propagate in any direction in the yz plane with the same x functional variation as given in Eq. (1.4) for TE modes and Eq. (1.25) for TM modes. For a planar guided-wave mode propagating in a direction θ with respect to the z axis, it will have a z variation of $\exp(-jn_{\mathrm{eff}}k(\cos\theta)z)$ and a y variation of $\exp(-jn_{\mathrm{eff}}k(\sin\theta)y)$. For such a planar guided wave, there is no variation of the field in the direction perpendicular both to x and to the direction of propagation.

There can be superposition of TE_m modes propagating in different θ directions to form diverging or focusing waves in the yz plane with identical x variation. Similarly, there can be superposition of TM_n modes propagating in different θ directions to form diverging or focusing waves in the yz plane with the same x variation. Notice that, for TE modes, the electric fields are polarized in the yz plane perpendicular to their direction of propagation; and the dominant magnetic field is polarized in the x direction. Vice versa, for TM modes, the magnetic fields are polarized in the yz plane perpendicular to their directions of propagation, while the dominant electric field is polarized in the x direction. When the waveguide has only a single mode in the x direction, there are only TE_0 modes in various directions of propagation in the yz plane. When both TE and TM modes exist in a given waveguide, there could be a mixture of TE and TM modes excited in a planar waveguide. Whether TE or TM modes will be excited depends on the polarization of the incident field.

In order to excite effectively a specific polarized guided-wave mode, the incident radiation must have a polarization close to the polarization of that mode. In a single mode waveguide, for incident radiation with polarization between the TE and TM polarization, only the TE mode will be excited by the component of the incident electric field polarized in the yz plane. The TM mode will be excited by the component of the incident electric field polarized in the x direction. When the waveguide has both TE and TM modes, TE modes will be excited by the component of the incident radiation with electric field polarized in the yz plane, while the TM modes will be excited by the component of the incident radiation with electric field polarized in the x direction. The direction of propagation of the guided-wave modes will be determined by the direction of the incident radiation beam through a relationship similar to Snell's law in free-space optics. Since TM and TE modes have different effective indices, they have different phase velocities. When both TE_0 and TM_0 modes are excited by a given incident radiation, the total polarization of the two modes will rotate as they propagate due to the difference in phase velocities.

In short, TE or TM modes of the same order all have the same $E_m(x)$ or $H_n(x)$ variation. There may be a number of planar guided waves with the same $E_m(x)$ or $H_n(x)$ simultaneously propagating in different θ directions in the yz plane. Superposition of such planar guided waves can give very complex field variations in the yz plane.

1.2.5.1 The Helmholtz equation for generalized guided-wave modes

We will now consider any generalized TE_m guided-wave mode in a planar waveguide to be a product $E_m(x)E_{m,t}(y,z)$. As long as the incident electric field is polarized in the yz plane, only the TE modes will be excited in the planar waveguide. In other words,

$$E_m(x, y, z) = E_m(x)E_{m,t}(y, z). \tag{1.38}$$

Note that $E_m(x)$ is the TE mth solution of Eq. (1.2) which has the eigenvalue β_m, or $n_{m,\text{eff}}k$. It is given in Eq. (1.4) as

$$\left[\frac{\partial^2}{\partial x^2} + \left(\omega^2 \mu \varepsilon(x) - n_{m,\text{eff}}^2 k^2 \right) \right] E_m(x) = 0,$$

$$\begin{aligned}
E_m(x) &= A_m \sin(h_m t + \phi_m) \exp[-p_m(x - t)] \quad &\text{for} \quad x \geq t \\
&= A_m \sin(h_m x + \phi_m) \quad &\text{for} \quad t > x > 0 \\
&= A_m \sin \phi_m \exp(q_m x). \quad &\text{for} \quad x < 0
\end{aligned}$$

Note that $E_{m,t}(y,z)$ is a function of y and z satisfying the two-dimensional scalar wave equation

$$\left(\frac{\partial^2}{\partial y^2} + \frac{\partial^2}{\partial z^2} + n_{m,\text{eff}}^2 k^2 \right) E_{m,t}(y, z) = 0. \tag{1.39}$$

When $E_{m,t}(y, z) = e^{-jn_{m,\text{eff}}k \sin \theta y} e^{-jn_{m,\text{eff}}k \cos \theta z}$, it is just the plane wave solution for $E_{m,t}(y,z)$ (i.e. a plane TE guided wave propagating in the yz plane in the θ direction). There are many other possible solutions. There is a strong similarity between the equation for $E_{m,t}(y,z)$ in Eq. (1.39) and the Helmholtz equation in optics. *All the techniques used to solve the Helmholtz equation can be applied here to $E_{m,t}(y,z)$. The major difference is that the Helmholtz equation in conventional optics is a scalar wave equation in three dimensions, while Eq. (1.39) is a scalar wave equation in two dimensions.* The mathematical details of how to solve scalar wave equations in two dimensions and in three dimensions are very different. A general method to solve Eq. (1.39) for a given incident radiation is the Green's function method [3]. Since guided waves are focused, collimated, and diffracted in many planar waveguide devices, the analysis of $E_{m,t}$ is very important for calculating the diffraction of guided waves in planar waveguides. An example, using the Green's function for calculating the diffraction of TE_0 waves excited by an incident radiation over a finite aperture is given in [4].

Similar comments can be made for TM_n guided-wave modes when the incident magnetic field is polarized in the yz plane (or for an incident electric field polarized in the x direction). In this case,

$$H_n(x,y,z) = H_n(x)H_{n,t}(y,z), \tag{1.40}$$

$$\left(\frac{\partial^2}{\partial y^2} + \frac{\partial^2}{\partial z^2} + n_{n,\mathrm{eff}}^2 k^2 \right) H_{n,t}(y,z) = 0, \tag{1.41}$$

where $H_n(x)$ is given in Eq. (1.25).

When the incident field has electric or magnetic field components polarized in both the x direction and a direction in the yz plane and when the waveguide can have both TE and TM modes, mixed TE and TM modes will be excited in the waveguide.

1.2.5.2   Examples of generalized guided waves in planar waveguides

In order to appreciate the importance of the more complex yz variation and the generalized guided-wave mode in planar waveguides, we will consider some examples using TE$_m$ modes in planar waveguides.

(A) Radiation from a line source in the yz plane

Let there be a single TE mode planar waveguide (i.e. the index and the thickness combination allows only the TE$_0$ mode to exist). A line source of guided-wave TE$_0$ mode is placed at the origin of the yz plane. A line source is represented mathematically as a unit impulse function δ in the yz plane. The solution for $E_{0,t}$ (y,z) in Eq. (1.39) is the cylindrical wave for distances far away from the origin,

$$E_{0,t}(y,z) = \frac{A}{\sqrt{\rho}} \mathrm{e}^{-jn_{0,\mathrm{eff}}k\rho}, \tag{1.42}$$

where

$$\rho = \sqrt{y^2 + z^2}. \tag{1.43}$$

Note that E approaches infinity as ρ approaches zero. This solution is similar to the spherical waves shown in conventional optics books except for the $1/\sqrt{\rho}$ variation instead of the $1/R$ variation in spherical waves. This modification is necessary if we consider the power P radiated by such a cylindrical wave in the yz plane in the form of a TE guided wave

$$P = \left\{ \frac{n_{\mathrm{eff},0}k}{2\omega\mu} \int_{-\infty}^{\infty} |E_0(x)|^2 \mathrm{d}x \right\} \int_{-\pi}^{\pi} |E_{0,t}(y,z)|^2 \rho \sin\theta \, \mathrm{d}\theta$$

$$= 2\pi A^2. \tag{1.44}$$

In evaluating P, we already knew from Eq. (1.11) that the result of the integration in x in { } is 1. Therefore P becomes proportional to A^2. In other words, the square root dependence in ρ is necessary for power conservation, i.e. for P to be independent of ρ. Notice also that the $E_{0,t}$ in Eq. (1.42) satisfies Eq. (1.41) only for large ρ when higher orders of $1/\rho$ can be neglected.

(B) The cylindrical guided-wave lens

Similarly to the three-dimensional case, an ideal guided-wave cylindrical lens at $z = 0$, placed parallel to the xy plane with the axis of the cylinder along the x axis, can be represented by a quadratic phase shift, $e^{j\frac{\pi n_{m,\mathrm{eff}}}{\lambda f}y^2}$. It will convert a planar guided wave normally incident on the lens into a convergent cylindrical guided wave focused at $z = f$. It will also collimate a divergent guided wave into a collimated guided wave.

In practice, it is difficult to obtain waveguide structures such that the effective index of the guided-wave mode within the lens is much larger than the effective index of the guided wave outside the lens. This difficulty is similar to that in making a three-dimensional lens with materials such that their index is not much larger than the index of air. Such lenses will be very weak. For these reasons, Fresnel lenses and Geodesic lenses are usually used for collimating and focusing guided waves in planar waveguides [5].

1.2.6   Rectangular channel waveguides and the effective index analysis

Rectangular waveguides are important in many practical applications because the rectangular cross-section is an idealized cross-section of actual waveguides fabricated by most micro-fabrication procedures such as etching. Figure 1.7 illustrates the index profile of two rectangular channel waveguides. In each case, the center portion, at $W/2 \geq |Y|$, consists of a ridge with a finite width W. Because of the complexity of the geometry of the dielectric boundaries, there is no analytical solution of the modes of such a structure. There are only approximate solutions [6] and computer programs such as Rsoft BeamProp© that can calculate numerically the guided-wave modes. These computer programs use numerical methods such as the beam propagation method or the finite element method for simulation [7]. The guided-wave modes could also be obtained easily by an approximation method called the effective index method, which will be presented here. This method is

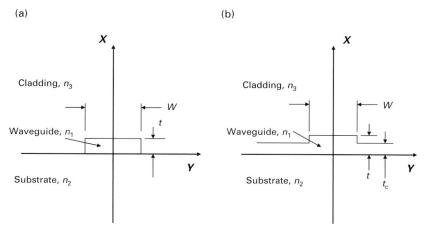

Fig. 1.7.   The index profile of two examples of channel waveguides. (a) Cross-sectional view of an etched channel waveguide. (b) Cross-sectional view of an etched ridged channel waveguide.

reasonably accurate for strongly guided modes (i.e. modes well above cut-off). It is based on the solutions of the planar guided-wave modes discussed in Sections 1.2.3 and 1.2.4. The effective index analysis also provides much insight into the properties of channel guided-wave modes.

Let us consider the rectangular channel waveguides in Fig. 1.7(a) and Fig. 1.7(b) where there is a rectangular core region, $y \leq |W/2|$, and a cladding region, $y \geq |W/2|$. Let us also assume that the planar waveguide in the core has only one mode in the x direction, the TE_0 mode. The propagation of the TE_0 planar guided-wave mode in the x direction in the core along its longitudinal direction z is given by $\exp(\pm j\beta_0 z)$ where β_0/k is its effective index, n_{e1}. In Fig. 1.7(b), there is also a planar guided-wave mode in the cladding region where we do not have the ridge. Let the effective index of the TE_0 planar guided-wave mode of the structure in the cladding region be n_{e2}. Since the high index layer is thicker for $y \leq |W/2|$, $n_1 > n_{e1} > n_{e2} > n_2$. In Fig. 1.7(a), there is no guided-wave mode in the cladding region, there are only continuous substrate and air modes for $y \geq |W/2|$. These continuous modes will have propagation wave number β in the x direction, where $n_{e1} > n_2 > \beta/k > n_3$.

Let us now consider the channel waveguide in Fig. 1.7(b). The core planar guided-wave mode in the $y \leq |W/2|$ region can propagate in any direction in the yz plane. Let us consider a core planar guided wave propagating in a direction which makes a very small angle δ with respect to the z axis. Let δ be so small that $n_{e1} \cos \delta > n_{e2}$. When this core planar guided wave is incident on the vertical boundary at $y = |W/2|$, it excites the cladding guided-wave mode at $y > |W/2|$ plus continuous modes. However, in order to match the boundary condition at $y = |W/2|$ as a function of z, the cladding guided-wave mode cannot have a real propagating wave number in the y direction. It must have an exponentially decaying y variation. In other words, the core planar guided wave is now totally internally reflected back and forth between the two boundaries at $y = \pm W/2$. The sum of all the reflected core planar guided waves would yield a non-zero solution only when the round trip phase shift of total internal reflection at specific values of δ is a multiple of 2π. These special sets of totally internally reflected core planar waveguide modes constitute the channel guided-wave modes. The lowest order mode (i.e. the 0th order) in the y direction has a round trip phase shift of 2π, and the nth order mode has a round trip phase shift of $2(n+1)\pi$. Consequently the field of the 0th order mode has no node in the y direction in the core, while the nth order mode has n nodes. If W is not very large, then we would also expect to have approximately a single mode in the y direction in the core. We call this mode the TE_{00} mode.

Consider now the mathematical details discussed in the preceding paragraph. At the $y = |W/2|$ boundaries, the electric field E_y of the core planar guided-wave mode is no longer the field transverse to the boundaries. Note that E_y is now approximately perpendicular to the boundaries, which are the $y = \pm W/2$ planes. The magnetic field has two components, H_x and H_z. The dominant tangential field of the core planar guided wave is H_x. Therefore, at the $y = |W/2|$ boundary, we will match the magnetic field H_x of the core and cladding modes.

The transverse field in the cladding that matches closely to the x variation of H_x of the guided-wave mode in the core at the $y = |W/2|$ boundary, is the sum of the H_x of the cladding TE planar guided wave of the same order and radiation modes. Since the cladding TE planar guided-wave mode has a similar x variation as the TE guided-wave mode in the core, it is the dominant component of the field in the cladding near the boundary. In order to satisfy the boundary condition for all z values, the z variation of this cladding guided-wave mode must be equal to $\exp(-jn_{e1}\cos\delta)$. If we let the y variation of the cladding guided wave be $\exp(-j\gamma y)$, γ must satisfy the equation

$$\gamma^2 = n_{e2}^2 - n_{e1}^2 \cos^2\delta. \tag{1.45}$$

Thus γ is imaginary when $n_{e1}\cos\delta > n_{e2}$. An imaginary γ represents an exponentially decaying cladding guided wave in the y direction, not a propagating cladding guided wave. In other words the core guided wave is totally internally reflected back at the $y = |W/2|$ boundaries. The nth channel guided-wave mode is obtained by demanding that the total round trip phase shift (with total internal reflection at the $y = \pm W/2$ boundaries) of the core planar guided wave (at angle δ) be $2(n+1)\pi$. If W is not very large, then we would also expect to have only a single mode in the y direction in the core. We call this mode the TE_{00} mode.

In short, the mathematics used here for analyzing the total internal reflection of the core planar guided wave in the y direction is equivalent to analyzing the total reflection of the equivalent TM planar wave propagating in the yz plane at angle δ with the magnetic field polarized approximately in the x direction. The equivalent material refractive indices are n_{e1} and n_{e2}, and the magnetic field is the transverse field. In other words, we can use the TM planar guided-wave mode equation for a symmetric waveguide in the y direction, i.e.

$$\left[\frac{\partial^2}{\partial z^2} + \frac{\partial^2}{\partial y^2} + \omega^2 \varepsilon(y)\mu\right] H_x(y,z) = 0,$$

$$\varepsilon(y) = \varepsilon_0 n_{ej}^2, \qquad j = 1 \text{ or } 2$$

$$\frac{\partial}{\partial x} = 0,$$

$$E_y = \frac{j}{\omega\varepsilon(y)}\frac{\partial H_x}{\partial z} \quad \text{and} \quad E_z = \frac{-j}{\omega\varepsilon(y)}\frac{\partial H_x}{\partial y}.$$

The boundary conditions are the continuity of H_x and E_z at $y = \pm W/2$. The nth solution of this equation will yield the effective index and the y variation of channel guided-wave mode TE_{0n} that we are looking for.[8] This is the effective index method.

It is important to use the TM equation because the field tangential to the $y = |W/2|$ plane is the magnetic field. The most important quantity to be obtained is the effective index, i.e. the β_n/k or $n_{e1}\cos\delta_n$, of the channel waveguide in the z direction. Knowing this effective index, we know both the δ in the core and the exponential decay constant, γ, in the cladding. Since δ is very small, the channel guided-wave mode obtained from the TE core planar guided mode is still approximately a y polarized TE mode propagating in the z direction. Naturally the x variation of E_y for $y < |W/2|$ is approximately the same as for the core planar guided-wave mode TE_0.

Notice that we no longer have pure TE or TM modes. We have basically TE-like modes with an electric field polarized in the y direction and a small electric field component in the z direction. These modes are called hybrid modes. Note also that the effective index approximation did not give us a solution for the x variation of the electric field near the boundaries. The boundary conditions at $y = \pm W/2$ are not satisfied by just the core and the cladding guided waves. In order to satisfy the boundary conditions accurately, many other modes, especially the substrate and air modes, must be involved. Air and substrate modes will decay in the y direction at different rates than the γ given in Eq. (1.45). However, based on the knowledge of continuity of the tangential electric field, we expect the electric and magnetic field to be continuous in the vicinity of $y = \pm W/2$. Thus, it is customary to assume that, near the boundaries, the field has the same x variation as the field in the core with an exponential decay rate of γ in the y direction.

For the waveguide shown in Fig. 1.7(a), the x variation of the tangential field of the core guided wave propagating at angle δ is matched by the summation of the continuous cladding modes at $y = |W/2|$. Since $n_{e1} \cos \delta > n_3$ and n_2, in order to satisfy the boundary condition as a function of z, all continuous modes will decay exponentially away from the $y = |W/2|$ boundary. Thus the core guided-wave mode is again totally internally reflected back and forth. The sum of all the reflected core planar guided waves would yield a non-zero solution when the round trip phase shift of total internal reflection for specific values of δ is a multiple of 2π. These special sets of totally internally reflected core planar waveguide modes constitute the channel guided-wave modes. However, in this case, we know the n_{e1} of the core TE planar guided-wave mode, but we do not know n_{e2} of the dominant field outside the core. Since a combination of substrates and air modes is used to match the x variation of the core guided wave at $y = \pm |W/2|$, the value of n_{e2} is somewhere between n_3 and the substrate index n_2. The best effective index n_{e2} to be used for the cladding region in the TM equation in y will depend on the profile of the core TE mode. For a high index waveguide with deep side walls, we will most likely use n_3 of the cladding. For a core guided wave with a long evanescent tail in the x direction in the substrate, we may use the substrate index. Fortunately, for well-guided channel modes in the core, the solution of n_{eff} and the y variation is not very sensitive to the value of the n_{e2} used for the calculation. Clearly, the approximation of the effective index method is not very good for such a structure. It is even more difficult to say anything about the x variation of the field in the cladding. The best we can do is to estimate the γ in the cladding region and to assume that for $|y| - |W/2| \ll \gamma$, the x variation is similar to the core guided-wave mode.

Similarly, a channel guided-wave mode with approximately TM polarization can be obtained from TM planar guided-wave modes in the core and in the cladding region. In that case the equivalent TE guided-wave equation will be used to find the effective index of the channel waveguide mode and the y variation.

1.2.6.1   An example of the effective index method

Consider first a GaAs planar waveguide with $n_1 = 3.27$ and $n_2 = 3.19$ and $t = 0.9$ μm in the core region operating at $\lambda = 1.5$ μm. This waveguide is exposed to air with $n_3 = 1$. The GaAs layer has been partially etched away at $y \geq |W/2|$, $W = 3$ μm. In the lateral cladding region, $t = 0.6$ μm. We would like to find the effective index and the field of the lowest order TE-like channel waveguide mode.

The first step of our calculation is to find the effective index of the TE_0 planar guided wave in the core region at $W/2 \geq |y|$ and in the cladding region at $|y| > W/2$. From Eq. (1.4) and (1.5), we find the TE planar guided-wave modes for the core and the cladding regions, $n_{e1} = 3.223$ and $n_{e2} = 3.211$. In accordance with Section 1.2.4.2, we solve the following equations to obtain the lowest order channel waveguide mode in the y direction (i.e. $n = 0$):

$$\tan\left[(h'_n/k)\frac{kW}{2}\right] = \frac{n_{e1}^2 p'_n/k}{n_{e2}^2 h'_n/k}, \qquad \left(\frac{h'_n}{k}\right)^2 + \left(\frac{p'_n}{k}\right)^2 = n_{e1}^2 - n_{e2}^2. \tag{1.46}$$

The solution is $(h'_0/k) = 0.1795$, which gives $n_{\text{eff},0} = 3.218$ and $p'_0/k = 0.2121$. The field distributions are approximately

$$
\begin{aligned}
E_y &= A \sin(h_0 x + \phi_0)\, \sin(h'_0 y + \phi'_0) e^{-jn_{\text{eff},0}kz} && \text{for } 0 < x \leq t,\ y \leq |W/2| \\
&= A \sin\phi_0 e^{q_0 x} \sin(h'_0 y + \phi'_0) e^{-jn_{\text{eff},0}kz} && \text{for } x \leq 0,\ y \leq |W/2| \\
&= A \sin(h_0 t + \phi_0) e^{-p_0(x-t)} \sin(h'_0 y + \phi'_0) e^{-jn_{\text{eff},0}kz} && \text{for } x \geq t,\ y \leq |W/2| \\
&= A \sin(h_0 x + \phi_0) \sin\left(\frac{h'_0 W}{2} + \phi'_0\right) e^{-p'_0\left(y-\frac{W}{2}\right)} e^{-jn_{\text{eff},0}kz} && \text{for } 0 < x < t,\ y > W/2 \\
&= A \sin(h_0 x + \phi_0) \sin\left(-\frac{h'_0 W}{2} + \phi'_0\right) e^{p'_0\left(y+\frac{W}{2}\right)} e^{-jn_{\text{eff},0}kz} && \text{for } 0 < x < t,\ y < -W/2 \\
&= A \sin(h_0 t + \phi_0) e^{-p_0(x-t)} \sin\left(\frac{h'_0 W}{2} + \phi'_0\right) e^{-p'_0\left(y-\frac{W}{2}\right)} e^{-jn_{\text{eff},0}kz} && \text{for } x > t,\ y > W/2 \\
&= A \sin(h_0 t + \phi_0) e^{q_0 x} \sin\left(\frac{h'_0 W}{2} + \phi'_0\right) e^{-p'_0\left(y-\frac{W}{2}\right)} e^{-jn_{\text{eff},0}kz} && \text{for } x < 0,\ y > W/2 \\
&= A \sin(h_0 t + \phi_0) e^{-p_0(x-t)} \sin\left(\frac{h'_0 W}{2} + \phi'_0\right) e^{p'_0\left(y+\frac{W}{2}\right)} e^{-jn_{\text{eff},0}kz} && \text{for } x > t,\ y < -W/2 \\
&= A \sin(h_0 t + \phi_0) e^{q_0 x} \sin\left(\frac{h'_0 W}{2} + \phi_0\right) e^{-p'_0\left(y+\frac{W}{2}\right)} e^{-jn_{\text{eff},0}kz}. && \text{for } x < 0,\ y < -W/W22
\end{aligned}
$$

$$\tag{1.47}$$

Here ϕ_0, q_0, h_0 and p_0 are parameters of the planar guided-wave TE_0 mode in the core (given by Eq. (1.4) and (1.5) with $\beta_m = 3.223k$). Since we do not know in detail how the radiation modes vary, we cannot find the field distributions accurately in the regions ($x > t$, $|y| > W/2$) and ($x < 0$, $|y| > W/2$) from the effective index method. A reasonable estimation is that the fields decay exponentially in the y direction from the x variation of the field in the core at the y boundaries.

1.2.6.2 Properties of channel guided-wave modes

Channel waveguides are used mostly in guided-wave devices such as a directional coupler, Y-branch splitter, waveguide laser, guided-wave modulator, waveguide photodetector, waveguide demultiplexer, ring resonator, and waveguide filter. Properties of the channel guided-wave mode which are most important to these applications are the n_{eff}, the attenuation rate, the polarization of the mode, and the evanescent tails described by

p_m, q_m, and γ. Most active channel waveguide devices are a few centimeters or less in length. Thus, unlike optical fibers, any reasonable attenuation rate, such as a few dB/cm or less, may be acceptable in many practical applications. Active channel waveguide devices often involve one guided-wave mode interacting with another guided-wave mode. These interactions will be discussed in detail in the next chapter.

In principle, guided-wave modes of channel waveguides are hybrid modes, meaning that there are field components in all x, y, and z directions. However, from the effective index analysis, it is clear that guided-wave modes can be considered as total internal reflection of the TE$_m$ (or TM$_n$) planar guided-wave modes at the y boundaries at a very small propagation angle δ from the z axis. It means that the polarization of the TE modes still has predominantly an electric field in the y direction and a magnetic field in the x direction. Therefore they are still called TE$_{mn}$ (or TM$_{nm}$) modes.

1.2.7   The representation of fields and the excitation of guided-wave modes

Most commonly, the guided-wave mode (or modes) in an abruptly terminated planar or channel waveguide (or weakly guiding optical fiber) is excited at its end by incident radiation from a laser (or an abruptly terminated fiber). The second most common method is to excite the guided-wave mode by a phase-matched interaction of the incident radiation with the guided-wave mode, utilizing the evanescent tail of the mode in the lower index cladding. Examples include the directional coupler for channel waveguides or the prism coupler for planar waveguides. The excitation of a guided wave by coupled channel waveguides will be discussed in Chapter 2. The excitation of planar guided waves by a prism coupler will be discussed in Chapter 5. We describe here the end excitation of guided waves.

For analysis of end excitation, the electric (or magnetic) field incident on the end of the waveguide is represented in terms of the modes of the waveguide. Mathematically, modes of the waveguide are solutions of a second order differential equation with proper boundary conditions. Such solutions form a complete set. This means that any arbitrary incident field at the input end of the waveguide can be expanded as a summation of such a complete set of modes. The coefficient of expansion of a specific guided-wave mode is the magnitude of the mode excited by the incident radiation (i.e. the excitation efficiency). From the amplitude and phase of all the continuous modes, we can calculate the radiation field excited by the incident radiation. Since the modes of planar and channel waveguides are divided into TE and TM sets, the transverse incident field will be first divided into x and y polarization components. The y component of the incident electric field will be expressed as a summation of TE modes, while the x component will be expressed as a summation of TM modes.

In order to understand more thoroughly the representation of the field and the excitation process, let us consider a specific example. Let the waveguide be excited by laser radiation from $z < 0$. The laser radiation and the waveguide are all oriented along the z direction. Let the waveguide at $z > 0$ be abruptly terminated at $z = 0$. At $z = 0$, both the x and y polarized electric fields must be continuous across the $z = 0$ plane. The transverse electric field of the incident radiation may be expressed as

$$\underline{E_t}(x,y) = E_x(x,y)\underline{i_x} + E_y(x,y)\underline{i_y}, \tag{1.48}$$

where $\underline{i_x}$ and $\underline{i_y}$ are unit vectors in the x and y directions. For $z \leq 0$, $\underline{E_t}$ consists of the incident laser mode and the reflected and diffracted laser radiation. If we neglect the reflection and diffraction at $z = 0$, $\underline{E_t}$ is just the incident laser radiation. For $z \geq 0$, $\underline{E_t}$ consists of the guided-wave modes and radiation (or cladding) modes of the waveguide. At $z = 0$

$$E_x = \sum_j A_j\psi_{x,j}(x,y) + \int_\beta b(\beta)\psi_x(\beta;x,y)\mathrm{d}\beta, \tag{1.49}$$

and

$$E_y = \sum_j C_j\psi_{y,j}(x,y) + \int_\beta d(\beta)\psi_y(\beta;x,y)\mathrm{d}\beta. \tag{1.50}$$

Here $\psi_{x,j}$ is the jth x polarized guided wave mode (i.e. the TM_{mn} modes), $\psi_{y,j}$ is the jth y polarized guided wave mode (i.e. the TE_{mn} modes), ψ_xs are the x polarized radiation modes, and ψ_ys are the y polarized radiation modes. Since the modes are orthogonal to each other (or non-overlapping), we can multiply both sides of Eq. (1.49) by $\psi_{x,j}(x,y)$ and integrate with respect to x and y from $-\infty$ to $+\infty$. In that case, we obtain:

$$|A_j|^2 = \frac{\left|\int\limits_{-\infty}^{\infty} \mathrm{d}x \int\limits_{-\infty}^{\infty} \mathrm{d}y \left[E_x\psi_{x,j}^*\right]\right|^2}{\left[\int\limits_{-\infty}^{\infty} \mathrm{d}x \int\limits_{-\infty}^{\infty} \mathrm{d}y |\psi_{x,j}|^2\right]^2}. \tag{1.51}$$

The expression, $\int\limits_{-\infty}^{\infty} \int\limits_{-\infty}^{\infty} E_x\psi_{x,j}^* \mathrm{d}x\,\mathrm{d}y$, is called the overlap integral between the incident field and the jth order mode. Note that $|A_j|^2 \int\limits_{-\infty}^{\infty} \mathrm{d}x \int\limits_{-\infty}^{\infty} \mathrm{d}y |\psi_{x,j}|^2 / \int\limits_{-\infty}^{\infty} \mathrm{d}x \int\limits_{-\infty}^{\infty} \mathrm{d}y |E_t|^2$ is the power efficiency for coupling the laser radiation into the x polarized jth guided-wave mode. A similar expression is obtained for coupling into the y polarized guided-wave mode, $|C_j|^2$.

When guided modes in both polarizations are excited the total polarization and intensity pattern of the total radiation in waveguides will be position dependent because of the difference in the phase velocity of different modes. The intensity pattern is sensitive with respect to geometrical, strain or bending perturbations. Clearly, radiation (or cladding) modes are also excited at $z = 0$. However, they radiate away after a short propagation distance. Therefore, most of the time, only how the guided waves are excited is of practical interest.

The optical power density carried by all the modes is $\frac{1}{2}\mathrm{Re}\left(E_y x H_x^*\right)$ for TE polarization and $\frac{1}{2}\mathrm{Re}\left(E_x x H_y^*\right)$ for TM polarization where the total field is given by Eq. (1.49) and (1.50). The total power carried in the waveguide by all the modes is either $\frac{1}{2}\mathrm{Re}\left[\int\limits_{-\infty}^{+\infty} \int\limits_{-\infty}^{+\infty} E_y H_x^* \mathrm{d}x\mathrm{d}y\right]$ or $\frac{1}{2}\mathrm{Re}\left[\int\limits_{-\infty}^{+\infty} \int\limits_{-\infty}^{+\infty} E_x H_y^* \mathrm{d}x\mathrm{d}y\right]$. When modes are orthogonal the result of the integrations is simplified by Eq. (1.11) and (1.30) so that the total power carried by all the modes is just the sum of the power carried by each mode. However, the orthogonality condition does not apply to waveguides that have significant losses [2].

1.2.8   Scalar approximation of the wave equations for TE and TM modes

It is clear from the discussion in Sections 1.2.3 and 1.2.6 that, for TE modes propagating in the z direction, E_y and H_x are the dominant field. The equation for E_y can be written as:

$$E_y(x, y, z) = E_y(z)E_y(x, y),   \tag{1.52}$$

$$\frac{\partial^2}{\partial z^2} E_y(z) + n_{\text{eff}}^2 k^2 E_y(z) = 0,   \tag{1.53}$$

$$\left[\frac{\partial^2}{\partial x^2} + \frac{\partial^2}{\partial y^2} + \left(\omega^2 \mu \varepsilon(x, y) - n_{\text{eff}}^2 k^2\right)\right] E_y(x, y) = 0,   \tag{1.54}$$

where $\varepsilon(x,y)$ is given by the waveguide structure. The solution for $E_y(z)$ is $e^{\pm j n_{\text{eff}} kz}$. Therefore, E_y can also be calculated directly from Eq. (1.54) plus appropriate boundary conditions. The boundary conditions can be deduced from the boundary conditions used in the effective index approximation.

In accordance with the effective index approximation discussed in Section 1.2.6, we first obtain the x variation of E_y approximately from the solutions of the planar waveguide equations in the x direction in the core and in the cladding. The boundary condition in the x direction is the continuity of $E_y(x,y)$ and H_z (which is proportional to $\partial E_y/\partial x$ across the x boundaries).[9] The n_{eff} of the planar guided wave in the core is n_{e1}, and the n_{eff} of the planar guided wave in the cladding is n_{e2}. The channel waveguide mode variation in y is obtained by total reflection of the H_x planar wave at the y boundaries whenever the round trip phase shift is $2n\pi$ where n_{e1} and n_{e2} are used to represent core and cladding indices. In this calculation, the E_z is the transverse electric field that must be continuous across the y boundaries. The boundary conditions in the y direction are the continuity of H_x and E_z. Note that, for an H_x in the planar waveguide in the y direction, $E_z = \frac{-j}{\omega\varepsilon(y)}\frac{\partial H_x}{\partial y}$ and $E_y = \frac{\beta}{\omega\varepsilon(y)}H_x$. The continuity of H_x assures the continuity of D_y[10] which is proportional to $\varepsilon(y)E_y$ at the y boundaries. The continuity of E_z means the continuity of $\frac{1}{\varepsilon(y)}\frac{\partial H_x}{\partial y}$ at the y boundaries. Therefore, even when Eq. (1.54) is not solved by the effective index method, the boundary conditions are the continuity of E_y and $\partial E_y/\partial x$ at the x boundaries and continuity of $\varepsilon(y)E_y$ and $\frac{1}{\varepsilon(y)}\frac{\partial \varepsilon(y)E_y}{\partial y}$ at the y boundaries.

Similar comments can be made for TM channel waveguide modes using H_y. The TM modes are found by solving the equation

$$\left[\frac{\partial^2}{\partial x^2} + \frac{\partial^2}{\partial y^2} + \left(\omega^2 \mu \varepsilon(x, y) - n_{\text{eff}}^2 k^2\right)\right] H_y(x, y) = 0,   \tag{1.55}$$

where H_y and $1/\varepsilon(x)$ times $\partial H_y/\partial x$ are continuous across x boundaries while H_y and $\partial H_y/\partial y$ are continuous across y boundaries. Note that E_x is the dominant electric field.

1.3   Formation of optical waveguides

How optical waveguides are made depends entirely on the availability of materials and processing technology. The objective is to obtain a low loss waveguide with precise control of the effective index, evanescent tail and mode size. Choice of materials and fabrication technology is affected further by the need to use materials and structures that have effective electro-optic, electro-absorption, electro-refraction or carrier injection properties in order to achieve specific active device functions at specific wavelengths.

For planar waveguides, the high index film is often obtained by processes such as thermal evaporation, electron beam deposition or sputtering, diffusion, ion exchange, doping, chemical vapor deposition or epitaxial growth. Each process has its own advantages and disadvantages, and is applicable only to certain materials.

Many techniques used in fabricating planar waveguides are used first to obtain the desired index variation in the thickness direction, homogeneous films and smooth interface between different material layers before the fabrication of channel waveguides. For channel waveguides, the lateral index variation of the waveguide core is obtained most commonly by one of the following techniques.

(1) Etching (including wet chemical etching and dry reactive etching) of the material through a mask to form a ridge pattern.
(2) Diffusion or preferential ion exchange through a patterned source or a mask to give a higher index in the core.
(3) Photo refractive effect using a patterned or a scanned optic beam.
(4) Poling of electro-optic active polymers using a patterned electrode to increase the index.
(5) Epitaxial regrowth of a semiconductor that has a higher index in a channel etched in the lower index substrate to provide the core.

Photolithographic techniques such as etching or lift-off through a mask are the most commonly used methods to create the etched pattern of the ridge, the diffusion source, the exchange mask, or the pattern of the poling electrode. Photolithography determines the resolution and the roughness of the masks. In addition, selection of the chemicals for wet or dry etching and etching time controls the etching depth, vertical profile, and surface roughness of the waveguides. Since scattering is a major cause of attenuation in channel waveguides, the roughness created by photolithography processes is a major consideration in selecting the appropriate method to fabricate the channel waveguides. The resolution of the photolithography also limits the minimum width of channel waveguides that can be fabricated. The specific etching process may also create under or over etch, which creates trapezoid shaped side walls in the vertical direction. In semiconductors, selective etching and stop-etch layers may be used to get better etch-depth, roughness and vertical shape control.

The control of material indices and core size is most important in determining the effective index, the mode size, and the evanescent decay of the guided-wave mode. Note that a large mode size of single mode waveguides can be obtained by either having a large core with small index difference with the surrounding media or by having a long evanescent tail when the mode is near its cut-off. Waveguides with large cores that have substantially higher index than the surrounding material will have multimodes. Waveguides with small cores and large index difference with surrounding media will have a small mode. The attenuation of the guided-wave mode is caused mostly by scattering or by absorption. The homogeneity and uniformity of the core material determine the amount of volume scattering. The smoothness of the interface with the cladding region and the index discontinuity of the defects determine the surface scattering loss of the modes. The closer the mode near its cut-off, the larger is the scattering loss.

A thorough discussion of various technologies on different materials that could be used to form optical waveguides is beyond the scope of this book. Therefore only four examples of the formation of optical waveguides on $LiNbO_3$, InP (or GaAs), polymer materials, and Si will be discussed briefly here.

1.3.1   Formation of optical waveguides on $LiNbO_3$ substrates

Lithium niobate, $LiNbO_3$, is a piezoelectric single-crystal insulator. Many high speed guided-wave electro-optic modulators and switches are made from this material because of its large electro-optic coefficients. High quality x-cut or z-cut $LiNbO_3$ substrates several centimeters long are commercially available. It is very hard and difficult to etch. Very few materials with index higher than $LiNbO_3$ can be deposited on it and have good optical quality. Therefore, only two methods, Ti-diffusion and proton ion exchange, have been used successfully to create optical waveguides.

Lithium niobate is a uniaxial crystalline birefringent material in which the refractive index is different for the optical electric field polarized along the z axis (i.e. n_e) or the x and y axis (i.e. n_o) of the crystal. Therefore the TE and TM modes will have significantly different effective indices. Its electro-optic properties will be discussed later in Section 3.1.

For planar waveguides to be fabricated by diffusion, a Ti film is first deposited on the surface of x-cut or z-cut $LiNbO_3$ and then heated to 1173–1273 K under controlled vapor pressure of gases such as Ar, H_2O and LiO_2. Diffusion of Ti into $LiNbO_3$ causes an increase of both the extra-ordinary index n_e and the ordinary index n_o. The depth and the core index of the waveguides are controlled by the diffusion temperature and time. For channel waveguides, the Ti film is patterned first into a strip in the configuration of the waveguide by etching or lift-off before diffusion. The channel waveguide is formed in the pattern defined by the Ti pattern after diffusion. The scattering loss of $LiNbO_3$ waveguides depends on the composition of the ambient gases, the diffusion tube and the temperature profile of the diffusion process. Well made Ti-diffused waveguides have very low scattering loss. There is very little scattering loss or absorption. Thus, straight sections of Ti-diffused waveguides are expected to have an attenuation rate less than 0.1 dB/cm. Ti-diffused waveguides are graded index waveguides. The index variation produced by Ti-diffusion is described commonly by an error function or Gaussian profile [8, 9].

For ion exchange, x-cut or z-cut $LiNbO_3$ substrate is immersed into a molten bath of benzoic acid at some temperature between the melting point (395 K) and boiling point

(522 K) [10]. The exchange between Li$^+$ and H$^+$ ions increases the extra-ordinary index n_e, but not the ordinary index n_o. The depth of the waveguide is controlled by the exchange time. The waveguides so obtained have been shown to have a step like index profile. Channel waveguides are obtained by masking the areas surrounding the waveguide to prevent the exchange. Compared with Ti-diffused waveguides, the increase of extra-ordinary index is much greater in this process, and the processing temperature is much lower than the diffusion temperature. Therefore a tighter mode size can be obtained. However, the attenuation rate of ion exchanged waveguide is higher than that of the lowest attenuation rate of Ti-diffused waveguide. The absence of an increase of the ordinary index means that there is no TM guided-wave mode for x-cut samples and no TE mode for z-cut samples.

1.3.2 Formation of optical waveguides on GaAs and InP substrates

High quality and large area GaAs and InP substrates are available commercially. Successful growth of single crystalline material on GaAs and InP substrates by LPE (liquid phase epitaxy), VPE (vapor phase epitaxy), MOCVD (metal organic chemical vapor deposition), and MBE (molecular beam epitaxy) techniques has already been developed for some time. The favored technique for fabricating commercial electronic components is MOCVD because it can be used to grow films uniformly over large substrate areas, while MBE is very useful for research purposes with its flexibility. Films grown epitaxially by MOCVD or MBE have precise control of thickness. For example, quantum well hetero-junction structures have been grown by MBE and MOCVD. The index of the epitaxially grown layer can be controlled by its composition. The lower the bandgap of the material, the higher is its refractive index. The III-V compound semiconductors are attractive for optoelectronic devices because they also have reasonably high electro-optic coefficients (not as high as LiNbO$_3$), as well as electro-absorption and electro-refraction effects. These electro-optical effects will be discussed in Chapter 3. In addition, optical waveguide devices on GaAs and InP could potentially be integrated monolithically with electronic devices. Such a process is called optoelectronic integration. Optical waveguide modulators have already been integrated with semiconductor lasers, called photonic integration in the literature. Therefore, GaAs and InP waveguides are very important in optoelectronics.

These compounds, GaAs and InP, belong to the III-V group of semiconductors which have many alloy composition variations, ranging from AlAs and GaP to InAs, GaSb and InSb. These are single crystals with pseudo-cubic lattice symmetry, so they are optically isotropic ($n_e = n_o = n$). There are two important features of epitaxial growth for waveguide fabrication.

(1) In epitaxial growth, the alloy composition of the film can be different than that of the substrate, provided that the lattice constant is matched to the substrate.[11] Materials with different composition will have different bandgap energy. The higher the bandgap, the lower is the refractive index. Figure 1.8 shows the room temperature bandgap energy of III-V semiconductors as a function of the lattice constant. The

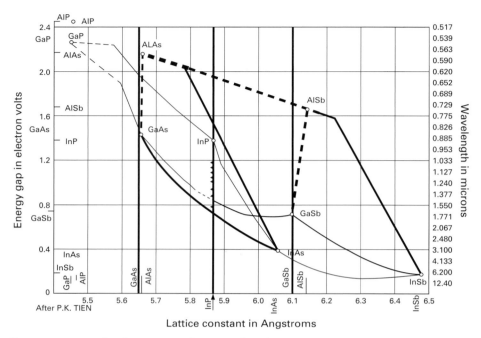

Fig. 1.8.   Room temperature bandgap energy of III-V semiconductor compounds vs. the lattice constant. Courtesy of Dr. P. K. Tien, AT & T Research Laboratory.

lattice constants of GaAs and InP are marked on the abscissa. There are many different alloy compositions that are matched to the lattice constant of GaAs or InP substrates. Therefore the growth of III-V semiconductor layers can provide us with high quality single-crystal layers that have precise bandgap or refractive index.

(2) The film thickness can be controlled very precisely with MOCVD or MBE growth, up to the precision of a single molecular layer. The surface of the epitaxially grown layer is atomically smooth.

In addition, semiconductors could have n-type or p-type doping, to form a p–i–n diode. In a reversed biased p-i-n diode, there is very little voltage drop in the p and the n layers. Most of the applied voltage is applied across the i layer. Since the thickness of the i layer is much smaller than the lateral dimension of an insulator waveguide such as LiNbO$_3$, we can obtain a high electric field in the i layer using a moderate electrical voltage. It means that large electro-optic or electro-absorption effects (to be discussed in Chapter 3) can be obtained with moderate applied voltage for modulation or switching.

Groups III-V semiconductors have a high refractive index, $n > 3$. Therefore surface roughness at the interface with any low index material such as the air will cause large scattering loss. Epitaxial growth provides us with an atomically smooth surface at the interfaces of the grown layers, it is good for planar waveguides. But for channel waveguides, etching is required to fabricate the lateral structure. Etching will generate surface roughness. Therefore, etched semiconductor waveguides like the one illustrated in Fig. 1.7(a) will have high attenuation. Most channel waveguides are formed by etching only a ridge in the cladding layer, similar to the one illustrated in Fig. 1.7(b), so that only the evanescent tail of the guided wave is scattered by the roughness of the etched surfaces. Alternatively, the channel waveguide can be fabricated by etch and regrowth. In this method, a channel is first etched into the host semiconductor. A lower bandgap material is regrown in the etched channel to form the higher index core. Surface roughness at the etched surface of the channel is smoothed out in the regrowth process.

1.3.3 Formation of polymer optical waveguides

Electro-optic polymer materials are amorphous organic materials composed of an electro-optically active component called a chromophore and a polymer matrix. For planar waveguides, the core layer is surrounded by lower index cladding (usually another polymer) layers above and below the core layer. The layers are usually spin coated and cured. The material is made electro-optically active by giving the ensemble of chromophore molecules an average alignment, usually by poling with an electric field at or above the glass temperature, T_g, and then cooling it to room temperature to freeze in the alignment. After poling, the material must not be heated to a temperature near the glass transition temperature, or the chromophore alignment will be lost. The attraction of polymer material is the very large electro-optic coefficient (much larger than those of $LiNbO_3$) that can be achieved in some engineered materials. It also has a refractive index only slightly higher than optical fibers as well as a low dielectric constant at microwave and radio frequencies (RF). Because of the small difference of the dielectric constants at optical and microwave frequencies, velocity matching can be easily accomplished in traveling wave devices for microwave and optical signals. Because of the low dielectric constant, microwave transmission lines will have higher impedances, easy to match. It is a material whose property depends on how it is synthesized. It can be spin coated on to a number of substrates such as glass [12].

A cross-section of a typical channel waveguide is shown in Fig. 1.9. The substrate is typically Si, chosen for its high surface quality and low cost. After depositing and

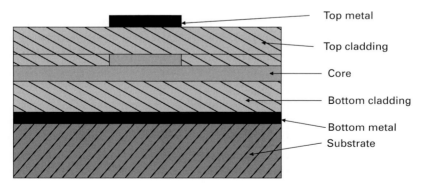

Fig. 1.9. An etched rib polymer waveguide. A shallow ridge is etched in the core layer, then it is covered with top cladding. The metal electrode could be used for poling and for applying the modulation electric field. (The figure is taken from ref. 12, Cambridge University Press.)

patterning of the bottom metal layer, the bottom cladding, the core, and the top cladding layer are deposited (or spin coated and cured) one at a time. The cladding material need not be electro-optic, but it needs to have a refractive index slightly less than that of the core. Finally the top metal layer is deposited and patterned. After fabrication of the layers, poling will be conducted at temperatures above the glass temperature by applying a voltage to the metal electrodes. The poling field is maintained while the temperature is lowered to 50–100 K below the glass temperature. At some point in the process, before the top metal pattern is deposited, one of the polymer layers is patterned to cause the lateral confinement needed to produce channel waveguides. In Fig. 1.9 the core layer was patterned. Methods of creating lateral optical confinement also include photobleaching and dry etching of the polymer core layer.

Notice that in order to take advantage of the very large electro-optic coefficient, the electric field needs to be polarized in the direction of the poling field. Properties of polymer waveguides depend a great deal on the material that can be synthesized. Ideally, a material should have high glass temperature, large electro-optic coefficient and low loss. But such an ideal material still remains to be developed.

1.3.4   Formation of optical waveguides on Si substrates

Since there are many commercial applications in Si electronics, very large Si wafers with superb surface quality can easily be obtained. Although Si is highly absorbing to visible light and near infrared radiation, it is transparent at a wavelength such as 1.55 μm. In addition, very high quality and thick SiO_2 buffer layers can be grown routinely on Si substrates by thermal oxidation or chemical vapor deposition to serve as substrates for waveguides. Alternatively, large area and high quality silicate substrates are also easily available. Various techniques such as ion exchange, doping and flame deposition can be used to obtain a higher index core to make waveguides. These waveguide materials are isotropic insulators. Such waveguides typically have very little scattering losses because of the superb interfacial quality and low index difference between the core and the cladding. Therefore optical waveguides on Si substrates have the lowest propagation loss. Because of their low loss, such waveguides are superb for applications such as ring resonators. However, these waveguides have no electro-optic or electro-absorption effect. Modulations of waves propagating in the waveguides can only be obtained through the absorption of free carriers injected into Si. No lasers or detectors can be photonically integrated with these waveguides. On the other hand, they can be optoelectronically integrated with Si electronics.

Planar waveguides have been fabricated by depositing SiO_2 buffer layer, phosphorus doped SiO_2 core layer, and SiO_2 cladding layer on Si substrate by plasma enhanced chemical vapor deposition [13]. The phosphosilicate glass core has 5–10% concentration of phosphorus while the buffer and cladding layers have 2–3% of phosphorus. The refractive index difference between the core and surroundings is controlled by the differ-ence in phosphorus concentration. The core layer was 5 μm thickness. The cladding and buffer layers are thicker. Channel waveguides were fabricated by first etching the core layer into a ridge, followed by deposition of the SiO_2 cladding on top. After deposition, the wafers were subjected to a 1273 K 3 hour annealing to reduce the O-H absorption in the 1–4 μm wavelength range. Losses in these channel waveguides were found to be about 0.1 dB/cm. Silver ion-exchanged channel waveguides were made by immersing a masked soda lime glass substrate into molten silver ions. The resulting lateral refractive index change is ~ 0.09 at the surface. The refractive index distribution may be modified by post-baking which relaxes the silver concentration gradient [14]. The single mode 2 μm wide waveguides made by this process were found to have 2 dB/cm loss.

Low loss waveguides have also been made by a combination of depositing high silica content glass on Si by flame hydrolysis and a reactive ion etching process. In this case, SiO_2 particles (a mixture of $SiCl_4$-$TiCl_4$ or $SiCl_4$-$GeCl_4$) for a buffer SiO_2 layer, followed by particles for a TiO_2-SiO_2 core layer, are first deposited on Si substrates by flame hydrolysis deposition. After deposition, the Si wafers with porous glass layers are heated to 1473–1573 K for consolidation. The desired SiO_2-TiO_2 core pattern for the channel waveguide is etched by reactive ion etching through a mask made from a sputtered Si material. Finally, a thick SiO_2 over-cladding layer on top of the core is formed by the same flame hydrolysis deposition process used to fabricate the buffer SiO_2 layer. The resultant channel waveguides have a relative index difference of 0.25–0.75 between the core and the surrounding materials, with a propagation loss from 0.1–0.3 dB/cm [15].

Notes

1. In differential equations, homogeneous solutions are solutions of the equations plus the boundary conditions without any source term.
2. Electric and magnetic fields tangential to the dielectric boundaries must be continuous.
3. Note that, according to Eq. (1.3), H_x is proportional to E_y while H_z is proportional to $\partial E_y/\partial x$. Therefore, continuity of $\partial E_y/\partial x$ is equivalent to the continuity of H_z which is parallel to the boundaries at $x = 0$ and $x = t$.
4. The plane waves in the cladding and substrate excited by the plane waves of the guided-wave mode have an imaginary propagation wave number in the x direction. Thus the guided wave has no radiation loss. In a different structure where a very low index layer is sandwiched between two high index media, there is also a solution of the plane waves reflected back and forth with 2π round trip phase shift. In this case, there is loss, produced by transmitted plane waves propagating away from the boundaries. These waves are not guided waves.
5. The modes are orthogonal only in lossless passive waveguides. When the media have absorption or amplification, modes may not be orthogonal [2].
6. In all lossless waveguides, radiation modes are orthogonal to guided-wave modes and each other [2].
7. Similarly to TE modes, the continuity of electric fields is equivalent to the continuity of H_y (i.e. $\varepsilon(x)E_x$) and $\frac{1}{\varepsilon(x)}\partial H_y/\partial x$ (i.e. E_z) across the x boundaries.
8. Note that the TE_{mn} channel waveguide mode is still polarized in the yz plane.
9. The dominant magnetic field of the TE planar waveguide mode transverse to the y boundaries is H_x. However, H_x is proportional to E_y in Maxwell's equations for a planar waveguide. Thus H_x is continuous as long as E_y is continuous.
10. Note that D_y is normal to the y boundaries.
11. Thin layers with slight lattice mismatch may also be grown epitaxially with accompanying strain. This technique may be useful to obtain material of a given bandgap with a composition that is not exactly matched to the substrate. However, its thickness is limited because the stress

creates defects. Too many defects may eventually cause material failure. However, using a step graded InAlAs buffer layer on GaAs substrate, Lei Shen was able to modify the lattice constants of the substrate. She has successfully grown InGaAs/InAlAs quantum well modulators on top of the buffer layer [11].

References

1. H. G. Unger, *Planar Optical Waveguides and Fibers*, Oxford University Press (1977).
2. D. Marcuse, *Light Transmission Optics*, Section 8.5, Van Nostrand Reinhold (1972).
3. P. M. Morse and H. Feshbach, *Methods of Theoretical Physics*, Section 7.2, McGraw-Hill (1953).
4. William S. C. Chang, *Principles of Lasers and Optics*, Chapter 3, Cambridge University Press, (2005).
5. M. C. Hamilton and A. E. Spezio, Spectrum analysis with integrated optics, Section 7.3.3 in *Guided-Wave Acousto-Optics*, ed. Chen S. Tsai, Springer-Verlag (1990).
6. D. Marcuse, *Theory of Dielectric Waveguides*, Chapter 1, Academic Press (1974).
7. R. Searmozzino, A. Gopinath, R. Pregla, and S. Helfert, Numerical techniques for modeling guided-wave photonic devices. *IEEE Select. Topics Quant. Electr.*, **6** (2000) 150.
8. J. P. Kaminow and J. R. Carruthers, Optical waveguiding layers in $LiNbO_3$ and $LiTiO_3$. *Appl. Phys. Lett.*, **22** (1973) 326.
9. G. J. Griffin, Optical waveguide fabrication techniques. Ph.D. thesis, University of Queensland, Australia (1981).
10. J. L. Jackel, C. E. Rice, and J. J. Vaselka, Proton exchange for high-index waveguides in $LiNbO_3$. *Appl. Phys. Lett.*, **41** (1982) 607.
11. Lei Shen, InGaAs/InAlAs quantum wells for 1.3 μm electro-absorption modulators on GaAs substrates. Ph.D. thesis, University of California, San Diego (1997).
12. T. Van Eck, Polymer modulators for RF photonics, Chapter 7 in *RF Photonic Technology in Optical Fiber Links*, ed. W. S. C. Chang, Cambridge University Press (2002).
13. G. Grand, J. P. Jadot, H. Denis, *et al.*, Low-loss PECVD silica channel waveguides for optical communications. *Electronics Lett.*, **26** (1990) 2136.
14. R. G. Walker, C. D. W. Wilkinson, and J. A. H. Wilkinson, Integrated optical waveguiding structures made by silver exchange in glass. 1: The propagation characteristics of stripe ion-exchanged waveguides; a theoretical and experimental investigation. *Appl. Optics*, **22** (1983) 1923.
15. Maso Kawachi, Silica waveguides on silicon and their application to integrated-optic components. *Opt. Quant. Elect.*, **22** (1990) 391.

2 Guided-wave interactions

The operation of many photonic devices is based on interactions between optical guided waves. We have discussed the electromagnetic analysis of the modes in individual planar and channel waveguides in Chapter 1. From that discussion, it is clear that solving Maxwell's equations rigorously for several coupled modes or waveguides is very difficult. Only approximate and numerical solutions are available. In this chapter, we will introduce several approximate electromagnetic techniques for analyzing the interactions of guided waves. These methods include the perturbation method and coupled mode analyses [1, 2]. Practical devices such as the grating filter, the directional coupler, the Y-branch coupler, the Mach–Zehnder modulator, and the multimode interference coupler will be discussed as specific examples. In addition, analysis of coupled waveguides as super modes of the total structure in the effective index approximation is presented. This analysis will allow us to view the interactions between coupled waveguides from another point of view.

In Chapter 1, we have shown that the guided-wave modes together with the radiation modes comprise a complete set of modes. In guided-wave devices, radiation modes are excited at any dielectric discontinuity. Rigorous modal analysis of propagation in a waveguide with varying cross-section in the direction of propagation should involve, in principle, all the modes. However, radiation modes usually fade away at some reasonable distance from the discontinuity. They are important only when radiation loss must be accounted for. Thus in the discussion of guided-wave interactions in this chapter, radiation modes such as the substrate and air modes in waveguides (and the cladding modes in fibers) are not included in our analysis. There are exceptions: for example, the radiation modes are very important in the analysis of a prism coupler in which a radiation beam excites a planar guided wave over a long interaction distance, or vice versa [3]. The prism coupler will be discussed in Chapter 5.

There are three types of guided-wave interaction which are the basis of the operation of most photonic devices.

(1) The adiabatic transition of guided-wave modes in waveguides (or fiber structures) in which the cross-section of the waveguides at one longitudinal position is transformed gradually to a different cross-section at another longitudinal position as the modes propagate. An example of this type of interaction is the symmetrical Y-branch that splits one channel waveguide into two identical channel waveguides. The combination of two symmetrical Y-branches with two well separated channel waveguides interconnecting them constitutes the well known Mach–Zehnder interferometer.

(2) The phase matched interaction between guided-wave modes of two waveguides over a specific interaction distance. A well known example of photonic devices based on this type of interaction is the directional coupler in channel waveguides (or fibers).
(3) Interaction of guided-wave modes through periodic perturbation of the optical waveguide. An example of this is the grating filter in channel waveguides (or optical fibers).

2.1 Perturbation analysis

Perturbation analysis is used to analyze the propagation of the guided wave in an optical waveguide as it is perturbed by another object in its vicinity.

2.1.1 Review of properties of modes in a waveguide

In any waveguide (or fiber) which has a transverse index variation independent of z (i.e. independent of the position along its longitudinal direction), the Maxwell's equations can be written in another form. The electric and magnetic fields, $\underline{E}(x,y,z)$ and $\underline{H}(x,y,z)$, propagating along the z axis, can be explicitly expressed in terms of the longitudinal ($\underline{E}_z,\underline{H}_z$) and transverse ($\underline{E}_t,\underline{H}_t$) fields as follows:

$$\underline{E} = \left[E_x\underline{i}_x + E_y\underline{i}_y\right] + E_z\underline{i}_z = \underline{E}_t + E_z\underline{i}_z = \underline{E}(x,y)\mathrm{e}^{-\mathrm{j}\beta z}\mathrm{e}^{\mathrm{j}\omega t},$$

$$\underline{H} = \left[H_x\underline{i}_x + H_y\underline{i}_y\right] + H_z\underline{i}_z = \underline{H}_t + H_z\underline{i}_z = \underline{H}(x,y)\mathrm{e}^{-\mathrm{j}\beta z}\mathrm{e}^{\mathrm{j}\omega t},$$

$$\nabla = \left[\frac{\partial}{\partial x}\underline{i}_x + \frac{\partial}{\partial y}\underline{i}_y\right] + \frac{\partial}{\partial z}\underline{i}_z = \nabla_t + \frac{\partial}{\partial z}\underline{i}_z,$$

$$\nabla_t \times \underline{E}_t = -\mathrm{j}\omega\mu H_z\underline{i}_z, \qquad \nabla_t \times \underline{H}_t = \mathrm{j}\omega\varepsilon(x,y)E_z\underline{i}_z,$$

$$\nabla_t \times E_z\underline{i}_z - \mathrm{j}\beta\underline{i}_z \times \underline{E}_t = -\mathrm{j}\omega\mu\underline{H}_t,$$

$$\nabla_t \times H_z\underline{i}_z - \mathrm{j}\beta\underline{i}_z \times \underline{H}_t = \mathrm{j}\omega\varepsilon(x,y)\underline{E}_t. \tag{2.1}$$

Equation (2.1) implies that the transverse fields can be obtained directly from the longitudinal fields, or vice versa, and either set specifies the field.

The nth guided-wave mode, given by $\underline{e}_n$ and $\underline{h}_n$, is the nth discrete eigenvalue solution of $\underline{E}$ and $\underline{H}$ in the above vector wave equation that also satisfies the condition of the continuity of tangential electric and magnetic fields across all boundaries. In view of the properties of the modes discussed in Chapter 1, we expect the following properties of the $\underline{e}_n$ and $\underline{h}_n$ modes for any general waveguide with constant cross-section in z.

(1) The magnitude of the fields outside the higher index core or channel region decays exponentially away from the high index region in lateral directions.
(2) The higher the order of the mode, the slower is the exponential decay rate.
(3) The effective index of the nth guided-wave mode $n_{\mathrm{eff},n}$ ($n_{\mathrm{eff},n} = \beta_n/k$) is less than the material index of the core and more than the material indices of the cladding and the substrate. Note that n_{eff} is larger for a lower order mode.

(4) Most importantly, it can be shown from the theory of differential equations, that the guided-wave modes of *lossless waveguides* are orthogonal to each other and to the substrate or cladding modes. Mathematically this is expressed for channel guided-wave modes as

$$\iint_S \left(\underline{e}_{t,m} \times \underline{h}_{t,n}^* \right) \cdot \underline{i}_z \mathrm{d}s = \int_{-\infty}^{\infty} \int_{-\infty}^{\infty} \left(\underline{e}_{t,m} \times \underline{h}_{t,n}^* \right) \cdot \underline{i}_z \mathrm{d}x \, \mathrm{d}y = 0, \quad \textit{for} \quad n \neq m, \qquad (2.2)$$

where the surface integral is carried out over the entire transverse cross-section with integration limits extending to $\pm\infty$. The guided-wave modes and all the radiation modes constitute a complete set of modes so that any field can be represented as a super-position of the modes. Moreover, the channel guided-wave modes are normalized, i.e.

$$\frac{1}{2} \mathrm{Re} \left[\iint_S \left(\underline{e}_{t,n} \times \underline{h}_{t,n}^* \right) \cdot \underline{i}_z \mathrm{d}S \right] = 1. \qquad (2.3)$$

For planar guided-wave modes, the modes are also orthogonal and normalized in the x variation as shown in Eq. (1.11) and (1.30). However, the integration in the y coordinate is absent. The normalization means that the power carried by the mth normalized planar guided-wave mode is one watt per unit distance (i.e. meter) in the y direction.

2.1.2   The effect of perturbation

Consider two waveguide structures that have the cross-sectional ε variation shown in Fig. 2.1(a) and Fig. 2.1(b). The original waveguide is shown in Fig. 2.1(a). The original waveguide plus perturbation is shown in Fig. 2.1(b). Let $\underline{E}$ and $\underline{H}$ be the solutions of Eq. (2.1) for the original waveguide with index profile $\varepsilon(x,y)$ shown in Fig. 2.1(a). Let $\underline{E}'$ and $\underline{H}'$ be the solutions of Eq. (2.1) for the waveguide structure with index profile $\varepsilon'(x,y)$ shown in Fig. 2.1(b). The two structures differ in the dielectric perturbation $\Delta\varepsilon$ shown in Fig. 2.1(c), where $\Delta\varepsilon(x,y) = \varepsilon'(x,y) - \varepsilon(x,y)$. Let us assume that $\underline{E}$, $\underline{H}$ and the

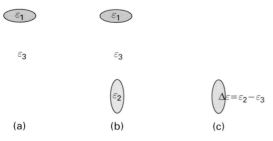

(a)   (b)   (c)

Fig. 2.1.   The index profile of a waveguide perturbed by $\Delta\varepsilon$. (a) The permittivity variation, $\varepsilon(x,y)$, of the original unperturbed waveguide structure. (b) The permittivity variation, $\varepsilon'(x,y)$, of the perturbed waveguide. (c) The permittivity perturbation from the additional material, $\Delta\varepsilon$, to the original waveguide structure.

guided-wave modes of the structure in Fig. 2.1(a) are already known. The guided-wave modes of the waveguide in Fig. 2.1(b) are the perturbation of the guided-wave modes of the structure in Fig. 2.1(a) due to the $\Delta\varepsilon$. The perturbation analysis allows us to calculate approximately the change of the $\underline{E}$ and $\underline{H}$ of the guided-wave modes without solving Maxwell's equations. Perturbation analysis is applicable as long as $\Delta\varepsilon$ is either small or at a position reasonably far away from the waveguide so that the evanescent tail of the mode for the original waveguide has decayed significantly.

Mathematically, from vector calculus and Eq. (2.1), we know

$$\nabla \cdot [\underline{E}^* \times \underline{H}' + \underline{E}' \times \underline{H}^*] = -j\omega \,\Delta\varepsilon \,\underline{E}^* \cdot \underline{E}'.$$

Let us apply volume integration to both sides of this equation over a cylindrical volume, V,

$$\iiint_V \nabla \cdot [\underline{E}^* \times \underline{H}' + \underline{E}' \times \underline{H}^*]\mathrm{d}x\,\mathrm{d}y\,\mathrm{d}z = -j\omega \iiint_V \Delta\varepsilon\underline{E}' \cdot \underline{E}^*\mathrm{d}x\,\mathrm{d}y\,\mathrm{d}z.$$

The cylinder has flat circular ends parallel to the xy plane. It has an infinitely large radius for the circular ends and a short length $\mathrm{d}z$ along the z axis. According to advanced calculus, the volume integration on the left hand side of this equation can be replaced by the surface integration of $[\underline{E}^* \times \underline{H}' + \underline{E}' \times \underline{H}^*]$ on the cylinder. The contribution of the surface integration over the cylindrical surface is zero because the guided-wave fields $\underline{E}$ and $\underline{E}'$ have already decayed to zero at the surface. For a sufficiently small $\mathrm{d}z$, $\underline{E}^* \cdot \underline{E}'$ is approximately a constant from z to $z + \mathrm{d}z$. Therefore, we obtain:

$$\iint_S \left\{ [\underline{E}^* \times \underline{H}' + \underline{E}' \times \underline{H}^*]|_{z+\mathrm{d}z} - [\underline{E}^* \times \underline{H}' + \underline{E}' \times \underline{H}^*]|_z \right\} \cdot \underline{i}_z \mathrm{d}S$$

$$= -j\omega \left[\iint_S \Delta\varepsilon \,\underline{E}' \cdot \underline{E}^*\mathrm{d}S \right] \mathrm{d}z.$$

Here S is the flat end surface of the cylinder oriented toward the $+z$ direction. In other words

$$\iint_S \frac{\partial}{\partial z} \left[\underline{E}_t^* \times \underline{H}_t' + \underline{E}_t' \times \underline{H}_t^* \right] \cdot \underline{i}_z \mathrm{d}S = -j\omega \iint_S \Delta\varepsilon(x,y)\underline{E}' \cdot \underline{E}^*\mathrm{d}S. \tag{2.4}$$

Mathematically, $\underline{E}'$ and $\underline{H}'$ can be represented by superposition of any set of modes. They can be either the modes of the structure shown in Fig. 2.1(b) or the modes of the structure shown in Fig. 2.1(a). Both sets of the modes, $(\underline{e}_{t,j}, \underline{h}_{t,j})$ and $(\underline{e}'_{tk}, \underline{h}'_{tk})$, form a complete orthogonal set. From the perturbation analysis point of view, we are not interested in the exact fields or modes of the structure shown in Fig. 2.1 (b). We only want to know how the fields for the waveguide in Fig. 2.1(a) are affected by $\Delta\varepsilon$.

In Eq. (2.4), let us express any $\underline{E}'$ and $\underline{H}'$ (in and near the waveguide with the core ε_1 and at any position z) in terms of the modes $(\underline{e}_{tj}, \underline{h}_{tj})$ as follows:

$$\underline{E}_t'(x, y, z) = \sum_j a_j(z)\underline{e}_{t,j}(x, y)\mathrm{e}^{-\mathrm{j}\beta_j z},$$

$$\underline{H}_t'(x, y, z) = \sum_j a_j(z)\underline{h}_{t,j}(x, y)\mathrm{e}^{-\mathrm{j}\beta_j z}. \tag{2.5}$$

The radiation modes have been neglected in Eq. (2.5). In general, the coefficients a_j may be different at different z. The variation of the a_j coefficient signifies how the $\underline{E}'$ and $\underline{H}'$ field may vary as a function of z. Substituting Eq. (2.5) into Eq. (2.4), letting $\underline{E}_t = \underline{e}_{t,n}$ and $\underline{H}_t = \underline{h}_{t,n}$, and utilizing the orthogonality and normalization relation in Eq. (2.2) and (2.3),[1] we obtain

$$\frac{\mathrm{d}a_n}{\mathrm{d}z} = -\mathrm{j} \sum_m a_m C_{m,n} \mathrm{e}^{+\mathrm{j}(\beta_n - \beta_m)z}$$

$$C_{m,n} = \frac{\omega}{4} \iint_S \Delta\varepsilon \left(\underline{e}_m \cdot \underline{e}_n^* \right) \mathrm{d}S. \tag{2.6}$$

This is the basic result of the perturbation analysis [4]. It tells us how to find the a_j coefficients. Once we know the a_j coefficients, we know $\underline{E}'$ and $\underline{H}'$ from Eq. (2.5) just for the region near the waveguide that has the ε_1 cross-section. We will apply this result to different situations in the next sections. Please note that the results shown in Eq. (2.6) do not tell us about the fields around the ε_2 in Fig. 2.1(b).

2.1.3   A simple application of perturbation analysis – perturbation by a nearby dielectric

In order to demonstrate the power of the results shown in Eq. (2.6), let us find the change in the propagation constant β_0 of a forward propagating guided-wave mode caused by the addition of another dielectric material with index ε' in the vicinity of the original waveguide. Let the original waveguide be located at $x = 0$ and $y = 0$. The dielectric material is located at $\infty > x \geq L$ and $\infty > y > -\infty$, $L > x$ dimension of the waveguide. Let us apply this $\Delta\varepsilon$ to Eq. (2.6). If the original waveguide has only a single mode, $\underline{e}_0$, then we do not need to carry out the summation in Eq. (2.6). We obtain

$$\frac{\mathrm{d}a_0}{\mathrm{d}z} = -\mathrm{j}a_0 \left[\frac{\omega}{4} \int_L^\infty \int_{-\infty}^\infty (\varepsilon' - \varepsilon_1)\underline{e}_0 \cdot \underline{e}_0^* \mathrm{d}x\,\mathrm{d}y \right] = -\mathrm{j}\Delta\beta\, a_0,$$

or

$$a_0 = A\,\mathrm{e}^{-\mathrm{j}\Delta\beta z}, \qquad \Delta\beta = \frac{\omega}{4}(\varepsilon' - \varepsilon) \int_t^{+\infty} \int_{-\infty}^{+\infty} \underline{e}_0 \cdot \underline{e}_0^* \,\mathrm{d}x\,\mathrm{d}y,$$

$$\underline{E}_t' = A\underline{e}_0(x, y)\mathrm{e}^{\mathrm{j}(\beta + \Delta\beta)z}. \tag{2.7}$$

Clearly the β_0 of the guided mode $\underline{e}_0$ is changed by the amount $\Delta\beta$. Notice that the perturbation analysis does not address the field distribution in the region $x > L$. The perturbation analysis allows us to calculate $\Delta\beta$ of the original waveguide mode without solving the differential equation.

2.2   Coupled mode analysis

2.2.1   Modes of two uncoupled parallel waveguides

Consider the two waveguides shown in Fig. 2.2(a). Let the distance of separation D between the two waveguides, A and B, be very large at first. In that case, the modes of A and B will not be affected by each other. The modes of the total structure, $\underline{e}_{tn}$ and $\underline{h}_{tn}$, are just the modes of individual waveguides, $(\underline{e}_{An}, \underline{h}_{An})$ and $(\underline{e}_{Bn}, \underline{h}_{Bn})$, or a linear combination of them. The fields of the total structure can be expressed as the summation of all the modes of the waveguides A and B

$$\underline{E} = \sum_n a_{An}\underline{e}_{An}e^{-j\beta_{An}z} + a_{Bn}\underline{e}_{Bn}e^{-j\beta_{Bn}z}$$

$$\underline{H} = \sum_n a_{An}\underline{h}_{An}e^{-j\beta_{An}z} + a_{Bn}\underline{h}_{Bn}e^{-j\beta_{Bn}z}. \qquad (2.8)$$

Here the "a" coefficients are independent of z. Because of the evanescent decay of the fields, the overlap of the fields $(\underline{e}_{An}, \underline{h}_{An})$ with $(\underline{e}_{Bn}, \underline{h}_{Bn})$ is negligible, i.e.

$$\iint\limits_{S} \left(\underline{e}_{t,An} \times \underline{h}_{t,Bm}^{*}\right) \cdot \underline{i}_z \mathrm{d}S = 0. \qquad (2.9)$$

In other words, A and B modes can be considered as orthogonal to each other.

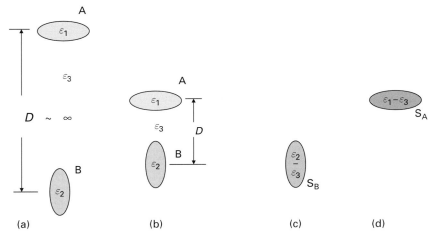

Fig. 2.2.   The mutual perturbation of two waveguides. (a) The permittivity profile of two well separated waveguides, A and B, with core dielectric constants ε_1 and ε_2. (b) The permittivity profile of two coupled waveguides, A and B, with core dielectric constants ε_1 and ε_2 separated by a moderate distance D. (c) The perturbation of ε_3 by ε_2 of the waveguide B for modes in the waveguide A. (d) The perturbation of ε_3 by ε_1 of the waveguide A for modes in the waveguide B.

2.2.2 Analysis of two coupled waveguides, using modes of individual waveguides

When the two waveguides are closer, but not very close to each other, the perturbed fields, $\underline{E}'$ and $\underline{H}'$, can again be expressed as a summation of ($\underline{e}_{An}$ and $\underline{e}_{Bn}$) and ($\underline{h}_{An}$ and $\underline{h}_{Bn}$) as follows:

$$\underline{E}' = \sum_n a_{An}(z)\underline{e}_{An}e^{-j\beta_{An}z} + a_{Bn}(z)\underline{e}_{Bn}e^{-j\beta_{Bn}z}$$

$$\underline{H}' = \sum_n a_{An}(z)\underline{h}_{An}e^{-j\beta_{An}z} + a_{Bn}(z)\underline{h}_{Bn}e^{-j\beta_{Bn}z}, \qquad (2.10)$$

where the "a" coefficients are now functions of z. However, the effect of the perturbation created by the finite separation distance D will be different for A and for B modes as shown below.

Consider now the two waveguides, A and B, separated by a finite distance D as shown in Fig. 2.2(b). For modes of waveguide A, the significant perturbation of the variation of the permittivity from the structure shown in Fig. 2.2(a) is the increase of permittivity from ε_3 to ε_2 at the position of the B waveguide as shown in Fig. 2.2(c). For modes of waveguide B, the perturbation of the variation of the permittivity is shown in Fig. 2.2(d) which is the increase of permittivity from ε_3 to ε_1 at the position of waveguide A. Applying the result in Eq. (2.6) to waveguides A and B separately, we obtain

$$\frac{da_{An}}{dz} = -j\left[C_{An,An}a_{An} + \sum_m C_{Bm,An}e^{j(\beta_{An}-\beta_{Bm})z}a_{Bm}\right],$$

$$\frac{da_{Bn}}{dz} = -j\left[C_{Bn,Bn}a_{Bn} + \sum_m C_{Am,Bn}e^{j(\beta_{Bn}-\beta_{Am})z}a_{Am}\right],$$

where,

$$C_{An,An} = \frac{\omega}{4}\iint_{S_B} (\varepsilon_2 - \varepsilon_3)\left[\underline{e}_{An} \cdot \underline{e}^*_{An}\right]dS,$$

$$C_{Bm,An} = \frac{\omega}{4}\iint_{S_B} (\varepsilon_2 - \varepsilon_3)\left[\underline{e}_{Bm} \cdot \underline{e}^*_{An}\right]dS,$$

$$C_{Bn,Bn} = \frac{\omega}{4}\iint_{S_A} (\varepsilon_1 - \varepsilon_3)\left[\underline{e}_{Bn} \cdot \underline{e}^*_{Bn}\right]dS,$$

$$C_{Am,Bn} = \frac{\omega}{4}\iint_{S_A} (\varepsilon_1 - \varepsilon_3)\left[\underline{e}_{Am} \cdot \underline{e}^*_{Bn}\right]dS, \qquad (2.11)$$

where the surface integration is carried out over the cylindrical flat end surfaces of waveguides A and B in Fig. 2.2.

Equation (2.11) is the well-known coupled mode equation [5]. It is used extensively to analyze many waveguide devices. There are a number of ways in which Eq. (2.11) may be simplified.

(1) Since there is evanescent decay of $\underline{e}_{An}$ before the field will reach S_B, $C_{An,An}$ is always much smaller than $C_{Bm,An}$. Similar comments can be made for $C_{Bn,Bn}$. Thus $C_{An,An}$ and $C_{Bn,Bn}$ are often neglected in Eq. (2.11) for a reasonably large separation distance D, specially when the effect on a_{An} and a_{Bn} by the $C_{Bm,An}$ and $C_{Am,Bn}$ is reasonably large. The example given in Section 2.1.3 illustrates the case when $C_{An,An}$ cannot be neglected.

(2) When there is no $\underline{e}_m$ mode in the second waveguide, $C_{Bm,An}$ or $C_{Am,Bn}$ will be zero, then $C_{An,An}$ is used to calculate the slight change of the propagation wave number of the modes, as we have done in Section 2.1.3.

(3) When there is more than one mode in waveguides A and B, there should also be more terms such as $C_{An,Aj}$ and $C_{Bn,Bj}$ in a more precise analysis. However, these C coefficients are even smaller than $C_{An,An}$ and $C_{Bn,Bn}$ because of the orthogonality of the unperturbed modes of the same waveguide.[2] Therefore, those terms have not been included in Eq. (2.11).

2.2.3   An example of coupled mode analysis – the grating reflection filter

Modes in different directions of propagation are independent solutions of the wave equations. For example, the independent modes can be the forward and backward propagating modes of the same order (or different orders) in a channel waveguide. They can be planar guided-wave modes in different directions of propagation in a planar waveguide. They can all be coupled by an appropriate $\Delta\varepsilon$ placed in the evanescent tail region. In the case of a prism coupler, there could even be the coupling of a guided-wave mode to substrate, air, or cladding modes (see Section 5.1.2.2). Equations (2.4) and (2.6) are directly applicable in analyzing such interactions. However, the details will differ for different applications. We will show in this section how the perturbation analysis could be used to analyze the coupling of modes in different directions of propagation via the grating reflection filter in an optical waveguide (or fiber).

Grating filters are very important devices in wavelength division multiplexed (WDM) optical fiber communication networks. In such networks, signals are transmitted via optical carriers that have slightly different wavelengths. The purpose of a filter is to select a specific optical carrier (or a group of optical carriers within a specific band of wavelength) to direct it (or them) to a specific direction of propagation (e.g. reflection) [6].

A grating reflection filter utilizes a perturbation of the channel waveguide by a periodic $\Delta\varepsilon$ to achieve the filtering function. The objectives of a grating filter are: (1) high and uniform reflection of incident waves in a waveguide within the selected wavelength band; (2) sharp reduction of reflectivity immediately outside the band; (3) high contrast ratio of the intensity of reflected optical carriers inside and outside the band.

Let us consider a grating layer which has a cosine variation of dielectric constant along the z direction, i.e. $\Delta\varepsilon(z)$, thickness d in the x direction and width W in the y direction. It is placed on top of a ridged channel waveguide that has a thickness t. An example of a ridged channel waveguide was shown in Fig. 1.7(b). Let us assume that the ridged waveguide has only a single mode.

Mathematically, let $\Delta\varepsilon = \Delta\varepsilon_0 \cos(Kz)\, \text{rect}\left(\frac{2(x-H)}{d}\right)\text{rect}\left(\frac{2y}{W}\right)$. It has a periodicity $T = 2\pi/K$ in the z direction and a maximum change of dielectric constant $\Delta\varepsilon_0$. The $\Delta\varepsilon$

perturbation layer is centered at $x = H$, where $H \geq t+(d/2)$. It is a perturbation of the cladding refractive index n_3 of the channel waveguide. This mathematical expression is a simplified $\Delta\varepsilon$ of a practical grating that normally has a $\Delta\varepsilon$ described by a rectangular function of x and z. Such a rectangular grating will be described in Section 5.2.1.

Let the complex amplitude of the forward propagating guided-wave mode be a_f and the amplitude of the backward propagating mode at the same wavelength be a_b. Then application of Eq. (2.4) to the field in the waveguide that has both the forward and the backward propagating modes yields

$$\underline{E}_t'(x, y, z) = \left[a_f(z)e^{-j\beta_0 z} + a_b(z)e^{+j\beta_0 z}\right]\underline{e}_{t,0}(x, y),$$

$$\frac{da_f}{dz} = -jC_{ff}a_f - jC_{bf}a_be^{-j2\beta_0 z},$$

$$\frac{da_b}{dz} = -jC_{bb}a_b - jC_{fb}a_fe^{j2\beta_0 z},$$

$$C_{ff} = -C_{bb} = -C_{fb} = C_{bf} = \frac{\omega}{4}\left[\int_{H-\frac{d}{2}}^{H+\frac{d}{2}}\int_{-\frac{W}{2}}^{\frac{W}{2}}\Delta\varepsilon_0\left|\underline{e}_0 \cdot \underline{e}_0^*\right|dx\,dy\right]\left[\frac{1}{2}\left(e^{jKz} + e^{-jKz}\right)\right], \qquad (2.12)$$

where there is a minus sign on C_{bb} and C_{fb}. Because, in the normalization of the modes shown in Eq. (2.3), the $\underline{i}_s$ is pointed toward the $+z$ direction. The $\underline{i}_z$ for the backward wave is pointing toward the $-z$ direction.

Clearly a_f and a_b will only affect each other significantly along the z direction when the driving terms on the right hand side of Eq. (2.12) have a slow z variation. Since the perturbation has a $\cos(Kz)$ variation, the maximum coupling between a_f and a_b will take place when $K = 2\beta_0$. This is known as the phase matching (or the Bragg) condition of the forward and backward propagating waves. When the Bragg condition is satisfied, the relationship between the βs and the K is illustrated in Fig. 2.3, where the β_0s of the forward and backward propagating modes with $\exp(\pm j\beta_0 z)$ variations are represented by vectors with magnitude β_0 in the $\pm z$ directions. Since a cosine function is the sum of two

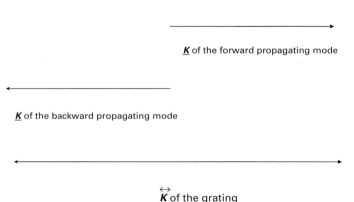

$\underline{K}$ of the forward propagating mode

$\underline{K}$ of the backward propagating mode

$\overleftrightarrow{K}$ of the grating

Fig. 2.3. Propagation wave vectors for forward and backward waves and the grating. The propagation wave vectors of the forward and backward guided waves are shown as $\underline{K}$ vectors in the $+z$ and $-z$ directions. The $\overleftrightarrow{K}$ of the grating is shown as a bi-directional vector. Phase matching is achieved when the magnitude of $|\overleftrightarrow{K}|$ is the sum of the $|\underline{K}|$.

exponential functions, $\underline{K}$ is represented as a bi-directional vector of magnitude K. If we designate λ_g as the free space wavelength in which the maximum coupling takes place, then the phase matching condition is satisfied when K is given by:

$$K = \frac{4\pi \, n_{\mathrm{eff}}}{\lambda_g}.$$ (2.13)

Here n_{eff} is the effective index of the guided-wave mode. When $K \cong 2\beta_0$, the terms involving C_{ff} and C_{bb} can be neglected in Eq. (2.12), in comparison with the terms involving C_{fb} and C_{bf}.

For a reflection filter, we like to have large a_{b} when any carrier frequency (i.e. β) is within the desired wavelength band. Since β is inversely proportional to λ, Eq. (2.12) will not be satisfied simultaneously for all the β within the desired band. In order to analyze the grating properties as a function of wavelength for a given K, we need to consider the solution of Eq. (2.12) under approximate phase matching conditions. Let

$$2\beta_0 - K = \delta_K.$$ (2.14)

Under this condition, we obtain from Eq. (2.12),

$$\frac{\mathrm{d}a_{\mathrm{f}}}{\mathrm{d}z} = -\mathrm{j}\frac{C_g}{2} a_{\mathrm{b}} \mathrm{e}^{\mathrm{j}\delta_K z},$$

and

$$\frac{\mathrm{d}a_{\mathrm{b}}}{\mathrm{d}z} = +\mathrm{j}\frac{C_g}{2} a_{\mathrm{f}} \mathrm{e}^{-\mathrm{j}\delta_K z},$$

where,

$$C_g = \frac{\omega}{4} \int\limits_{H-\frac{d}{2}}^{H+\frac{d}{2}} \int\limits_{-\frac{W}{2}}^{\frac{W}{2}} \Delta\varepsilon_0 \, |\underline{e}_{\mathrm{o}}|^2 \mathrm{d}x \, \mathrm{d}y.$$ (2.15)

Equation (2.15) is known as the coupled mode equation between the forward and the backward propagating modes. We know the solutions for such a differential equation are the familiar exponential functions, $\mathrm{e}^{\gamma^+ z}$ and $\mathrm{e}^{\gamma^- z}$. Specifically, the solutions of Eq. (2.15) for the forward and backward propagating waves are:

$$a_{\mathrm{b}}(z) = A_1 \mathrm{e}^{\gamma^+ z} + A_2 \mathrm{e}^{\gamma^- z},$$

$$a_{\mathrm{f}}(z) = -\mathrm{j}\frac{2}{C_g}\left[A_1\gamma^+ \mathrm{e}^{-\gamma^- z} + A_2\gamma^- \mathrm{e}^{-\gamma^+ z}\right],$$

$$\gamma^+ = -\mathrm{j}\frac{\delta_K}{2} + Q, \qquad \gamma^- = -\mathrm{j}\frac{\delta_K}{2} - Q,$$

$$Q = \sqrt{\left(\frac{C_g}{2}\right)^2 - \left(\frac{\delta_K}{2}\right)^2}.$$ (2.16)

The A_1 and A_2 coefficients will be determined from boundary conditions at $z = 0$ and $z = L$.

For a grating that begins at $z = 0$ and terminates at $z = L$, a_b must be zero at $z = L$. Thus

$$A_2 = -A_1 e^{2QL},$$

$$a_b = -A_1 2 e^{QL-j\left(\frac{\delta_K}{2}z\right)} \sinh[Q(L - z)],$$

$$a_f = -jA_1 \frac{4}{C_g} e^{QL+j\left(\frac{\delta_K}{2}z\right)} \left[j\frac{\delta_K}{2} \sinh(Q(L - z)) + Q \cosh(Q(L - z)) \right]. \tag{2.17}$$

At $z = 0$, the ratio of the reflected power to the incident power is

$$\frac{|a_b(z = 0)|^2}{|a_f(z = 0)|^2} = \frac{\left(\frac{C_g}{2}\right)^2 \sinh^2 QL}{Q^2 \cosh^2 QL + (\delta_K/2)^2 \sinh^2 QL}. \tag{2.18}$$

At $z = L$, the ratio of the transmitted power in the forward propagating mode to the incident power of the forward mode at $z = 0$ is

$$\frac{|a_f(z = L)|^2}{|a_f(z = 0)|^2} = \frac{Q^2}{Q^2 \cosh^2 QL + (\delta_K/2)^2 \sinh^2 QL}. \tag{2.19}$$

Since $|a_f(z = L)|^2 + |a_b(z = 0)|^2 = |a_f(z = 0)|^2$, the conservation of power of the incident, transmitted and reflected waves is verified. For a reflection filter, we want $|a_b(z = 0)/a_f(z = 0)|^2$ large within a desired band of wavelength, and small outside this band.

Notice that $|a_b(z=0)|$ is larger for larger L and smaller δ_K/C_g. At $\lambda = \lambda_g$, δ_K is 0, and the grating reflection is a maximum. The maximum possible value of $|a_b(z = 0)/a_f(z = 0)|^2$ is 1. At $\delta_K = C_g$, there will not be any reflected wave. Let $\Delta\lambda_g$ be the wavelength deviation from λ_g such that, when $\lambda = \lambda_g \pm \Delta\lambda_g$, Q is 0. Then $2\Delta\lambda_g$ is the pass band of the filter,

$$\Delta\lambda_g = \pm \frac{4\pi C_g n_{\text{eff}}}{K^2}. \tag{2.20}$$

In summary, K is used to control the center wavelength λ_g at which the transmission of the forward propagating wave is blocked. Note that C_g is used to control the wavelength width $\Delta\lambda_g$ within which effective reflection occurs. The smaller the C_g, the narrower the range of the transmission wavelength. For a given transmission range, L is used to control the magnitudes of the reflected and the transmitted wave. These are useful parameters for designing grating reflection filters.

2.2.4 An example of coupling of waveguides – the directional coupler

A directional coupler has an interaction region that has two parallel channel waveguides (or fibers). A prescribed fraction of power in waveguide A is transferred into waveguide B within the interaction region and vice versa. A top view of a channel waveguide directional coupler is illustrated in Fig. 2.4(a). Within the interaction region, the waveguides are separated from each other by a distance D, which is usually of the order of the evanescent decay length. Let the length of the interaction section be W. Outside the interaction region, the waveguides are well separated from each other without any further

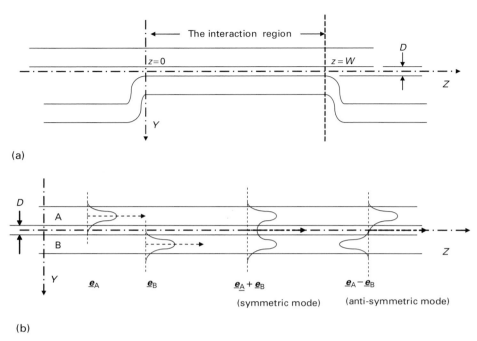

Fig. 2.4.   Top view of a directional coupler and illustration of coupled modes in the interaction region. (a) Top view of two channel waveguides in a directional coupler. The interaction region is W long. The separation distance of the two waveguides in the interaction region is D. (b) The field patterns of symmetric and anti-symmetric super modes of the two coupled identical waveguides in the interaction region. $\underline{e}_A$ and $\underline{e}_B$ are field patterns of the modes of the isolated waveguides A and B.

interaction. Clearly, Eq. (2.11) is directly applicable to the modes of the individual waveguides in the interaction region.

Let $\underline{e}_A$ and $\underline{e}_B$ be the modes of the two waveguides (or fibers) that are interacting with each other through their evanescent field in the interaction section, see Fig. 2.4(b). Let the two waveguides have cores with cross-sections, S_A and S_B, and dielectric constants, ε_A and ε_B. The cores are surrounded by a medium which has dielectric constant ε_3. Let the coupling region begin at $z = 0$ and end at $z = W$ as shown in Fig. 2.4(a). For mathematical convenience, the coupling is assumed to be uniform within this distance. Application of Eq. (2.11) yields

$$\frac{da_A}{dz} = -jC_{BA}e^{j\Delta\beta z}a_B(z),$$

$$\frac{da_B}{dz} = -jC_{AB}e^{-j\Delta\beta z}a_A(z),$$

$$C_{AB} = \frac{\omega}{4}\iint_{S_A}(\varepsilon_A - \varepsilon_3)\left[\underline{e}_A \cdot \underline{e}_B^*\right]dS,$$

$$C_{BA} = \frac{\omega}{4}\iint_{S_B}(\varepsilon_B - \varepsilon_3)\left[\underline{e}_B \cdot \underline{e}_A^*\right]dS,$$

$$\Delta\beta = \beta_A - \beta_B. \tag{2.21}$$

Here, C_{AA} and C_{BB} have been neglected in anticipation of the large effects to be produced by C_{AB} and C_{BA} at small $\Delta\beta$. Solution of a_A and a_B will depend again on initial conditions. Let the initial condition be $a_A = A$ and $a_B = 0$ at $z = 0$. Then, we obtain

$$
a_A = A e^{j\Delta\beta/2z} \left[\cos\left(\sqrt{C_{BA}C_{AB} + (\Delta\beta/2)^2} z \right) \right.
$$

$$
\left. - j \frac{(\Delta\beta/2)}{\sqrt{C_{BA}C_{AB} + (\Delta\beta/2)^2}} \sin\left(\sqrt{C_{BA}C_{AB} + (\Delta\beta/2)^2} z \right) \right],
$$

$$
a_B = \frac{-jC_{AB}A}{\sqrt{C_{BA}C_{AB} + (\Delta\beta/2)^2}} e^{-j\Delta\beta/2z} \sin\left(\sqrt{C_{BA}C_{AB} + (\Delta\beta/2)^2} z \right), \tag{2.22}
$$

for $0 \leq z \leq W$.

Similarly, if the boundary condition is $a_B = B$ and $a_A = 0$ at $z = 0$, we obtain:

$$
a_A = \frac{-jC_{BA}B}{\sqrt{C_{BA}C_{AB} + (\Delta\beta/2)^2}} e^{+j\Delta\beta/2z} \sin\left(\sqrt{C_{BA}C_{AB} + (\Delta\beta/2)^2} z \right),
$$

$$
a_B = B e^{-j\Delta\beta/2z} \left[\cos\left(\sqrt{C_{BA}C_{AB} + (\Delta\beta/2)^2} z \right) \right.
$$

$$
\left. + j \frac{(\Delta\beta/2)}{\sqrt{C_{BA}C_{AB} + (\Delta\beta/2)^2}} \sin\left(\sqrt{C_{BA}C_{AB} + (\Delta\beta/2)^2} z \right) \right], \tag{2.23}
$$

for $0 \leq z \leq W$.

At $z = W$, the power transmitted from one waveguide to another and the power remaining in the original waveguide are calculated from a_B and a_A. Note that, unless $\Delta\beta = 0$, there cannot be full transfer of power from A to B. Substantial transfer of power from A to B (or vice versa) at $z = W$ can take place only when $\Delta\beta$ is small. Note that $\beta_A = \beta_B$ is the phase matching condition for maximum transfer of power. Similarly to all coupled mode interactions, the C coefficients, the W and the $\Delta\beta$ are used to control the net power transfer from A to B and from B to A. If W is too large, then a_A and a_B will exhibit oscillatory amplitude as z progresses.

Conventionally, the directional coupler has two identical channel waveguides. In that case, $C_{BA} = C_{AB} = C$, and the ratio of $|a_B|^2/|a_A|^2$ is the power distribution between the two waveguides. At $z = 0$, let there be an input power I_{in} in waveguide A, no input power in waveguide B. Then the output power I_{out} in waveguide B after an interaction distance W is given directly by Eq. (2.22). It is

$$
I_{out}/I_{in} = \frac{1}{C^2 + \left(\frac{\Delta\beta}{2}\right)^2} \sin^2\left(\sqrt{C^2 + \left(\frac{\Delta\beta}{2}\right)^2} W \right). \tag{2.24}
$$

A directional coupler modulator is a directional coupler with electro-optical control of $\Delta\beta$.[3] Since the power transfer will be affected by $\Delta\beta$, it is an intensity modulator. Furthermore, the power transfer is dependent on the interaction length W.

The discussion presented in this section is also the approach used commonly in the literature to describe the directional coupler [7, 8]. However, the directional coupler can also be viewed as propagation of the super modes in the total two-waveguide structure in the following section. Such an approach has not been described in most optics books. The super mode analysis is very useful for understanding thoroughly devices such as Y-branch couplers that cannot be analyzed by coupled mode analysis.

2.3   Super mode analysis

The operation of a number of devices such as the directional coupler was analyzed in the previous section by perturbation analysis based on the mutual interactions of guided-wave modes of two parallel waveguides via the evanescent field. There is an alternative analysis of the operation of these devices based on the modal analysis of the total waveguide structure, called the super modes.

What is a super mode analysis? For infinitely long parallel waveguides with uniform cross-section and distance of separation, the modes of the total structure are called the super modes. Each mode has a different effective index. When more than one super mode is excited by the incident radiation, the total field pattern at different longitudinal positions will be given by the summation of all the super modes. *The super mode analysis is an analysis of waveguide devices based on the interference pattern of super modes. It is different from the coupled mode analysis because it does not assume that the modes of the individual waveguides are just perturbed by their neighbors. Therefore the super mode analysis is more accurate when the separation between waveguides is very small, or even zero. Devices analyzed by super mode analysis also shed different light in understanding the device operation.* We will present in the following subsections the super modes of two waveguides in more detail, followed by analyses of a directional coupler, a Y-branch coupler and a Mach–Zehnder interferometer as examples. An example of the directional coupler has already been discussed in Section 2.2.4 in terms of coupled mode analysis. It is interesting to compare the results of two different analyses. The Y-branch coupler is an example that cannot be analyzed by coupled mode analysis.

2.3.1   Super modes of two parallel waveguides

2.3.1.1   Super modes of two well separated waveguides

Consider the two waveguides shown in Fig. 2.2(a). Let the distance of separation G between the two waveguides, A and B, be very large at first. In that case, the modes of A and B will not be affected by each other. In other words, the fields of the total structure can be expressed as the summation of all the modes of the waveguides A and B,

$$\underline{E} = \sum_n a_{An}\underline{e}_{An}\mathrm{e}^{-\mathrm{j}\beta_{An}z} + a_{Bn}\underline{e}_{Bn}\mathrm{e}^{-\mathrm{j}\beta_{Bn}z}$$

$$\underline{H} = \sum_n a_{An}\underline{h}_{An}\mathrm{e}^{-\mathrm{j}\beta_{An}z} + a_{Bn}\underline{h}_{Bn}\mathrm{e}^{-\mathrm{j}\beta_{Bn}z}. \tag{2.25}$$

Here the "a" coefficients are independent of z. Since there is evanescent decay of the fields, the overlap of the fields ($\underline{e}_{An}$, $\underline{h}_{An}$) with ($\underline{e}_{Bn}$, $\underline{h}_{Bn}$) is negligible, i.e.:

$$\iint_S \left(\underline{e}_{t,An} \times \underline{h}^*_{t,Bm}\right) \cdot \underline{i}_z \mathrm{d}S = 0.$$

In other words, modes of A and B are considered to be orthogonal to each other. The super modes of the total structure, ($\underline{e}_{sn}$,$\underline{h}_{sn}$) and ($\underline{e}_{an}$,$\underline{h}_{an}$) are just linear combinations of the modes of individual waveguides, ($\underline{e}_{An}$,$\underline{h}_{An}$) and ($\underline{e}_{Bn}$,$\underline{h}_{Bn}$), such that

$$\underline{e}_{sn} = \frac{1}{\sqrt{2}}(\underline{e}_{An} + \underline{e}_{Bn}) \quad \text{and} \quad \underline{e}_{an} = \frac{1}{\sqrt{2}}(\underline{e}_{An} - \underline{e}_{Bn}). \tag{2.26}$$

When waveguide A is identical with waveguide B, these modes are the symmetric and anti-symmetric modes of the total structure. Note that although A and B are both single mode waveguides, there are still two modes for the total structure, $\underline{e}_{s0}$ and $\underline{e}_{a0}$.

2.3.1.2   Super modes of two coupled waveguides

When the distance of separation between the two waveguides is small, as shown in Fig. 2.2(b), we can use the effective index approximation or numerical methods to find the super modes. Consider two parallel channel waveguides, A and B, as depicted in Fig. 2.5. Figure 2.5(a) shows the cross-sectional view in the xy plane, while Fig. 2.5(b) shows the plan view in the yz plane. In this illustration, waveguide A has core thickness

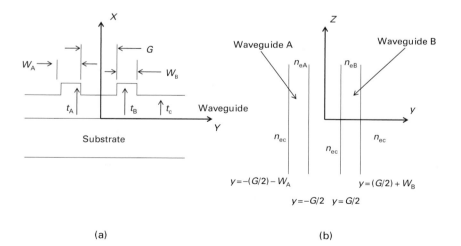

(a)   (b)

Fig. 2.5.   A two-channel-waveguide structure. Two parallel ridged waveguides with thicknesses, t_A and t_B, and widths, W_A and W_B are separated by a gap G, the cladding has thickness t_c. (a) Cross-sectional view. (b) Top view.

t_A and width W_A, while waveguide B has core thickness t_B and width W_B. The width of the gap between two waveguides is G. The thickness of the cladding is t_c. The substrate index is n_{sub}, while the index of the core of the waveguide and the cladding is n_{wg}. In accordance with the effective index method presented in Section 1.2.6, we first find the effective indices of the planar waveguide modes for the two waveguides and the cladding as we have done in Section 1.2.3.1. Let us assume that there is only a single TE_0 mode in the x direction. Let the effective index for the planar mode in waveguide A be n_{eA}; the effective index for the planar mode in waveguide B be n_{eB}; and the effective index for the planar mode in the cladding regions be n_{ec}. The lateral variation of the super mode is then found by solving the TM planar waveguide mode of H_x in the y direction,

$$\left[\frac{\partial^2}{\partial z^2} + \frac{\partial^2}{\partial y^2} + \omega^2 \varepsilon(y)\mu\right] H_x(y, z) = 0,$$

$$\varepsilon(y) = \varepsilon_0 n_{ej}^2, \qquad j = A, \, B \text{ or } c, \tag{2.27}$$

$$\frac{\partial}{\partial x} \equiv 0 \quad \text{for planar TM mode approximation,} \tag{2.28}$$

$$E_y = \frac{j}{\omega\varepsilon(y)}\frac{\partial H_x}{\partial z}, \tag{2.29}$$

$$E_z = \frac{-j}{\omega\varepsilon(y)}\frac{\partial H_x}{\partial y}, \tag{2.30}$$

where the boundary conditions are the continuity of H_x and E_z at $y = \pm|G/2|$ and $y = \pm|W + (G/2)|$.

2.3.1.3   An example: super modes of two parallel identical waveguides

When waveguides A and B are identical in Fig. 2.5, we let $n_{eA} = n_{eB} = n_1$, $W = W_A = W_B$, and $n_{ec} = n_2$. The super modes become the symmetric and anti-symmetric modes in the y direction. In the effective index approximation, the $E_y(x)$ of the planar waveguide modes are given in Eq. (1.4), while the effective indices are obtained from p_0 and q_0 in Eq. (1.5). The y variations of the super modes are:

(1) Symmetric mode:

$$H_x = B\cos\left[h_1\left(\frac{W}{2}\right) + \phi\right]e^{+q_2\left[y + \left(\frac{G}{2} + W\right)\right]}$$

$$\beta^2 - q_2^2 = k^2 n_2^2, \quad \text{for} \quad y \le -\frac{G}{2} - W \tag{2.31a}$$

$$H_x = B\cos\left[h_1\left(y + \frac{G+W}{2}\right) - \phi\right],$$

$$\text{for} \quad -\left(\frac{G}{2} + W\right) \le y \le -\frac{G}{2} \tag{2.31b}$$

$$H_x = B'[e^{-q_2 y} + e^{+q_2 y}]$$

$$\beta^2 + h_1^2 = k^2 n_1^2, \quad \text{for} \quad -\frac{G}{2} \leq y \leq +\frac{G}{2} \tag{2.31c}$$

$$H_x = B \cos\left[h_1\left(y - \frac{G+W}{2}\right) + \phi\right],$$

$$\text{for} \quad \frac{G}{2} \leq y \leq \frac{G}{2} + W \tag{2.31d}$$

$$H_x = B \cos\left[h_1 \frac{W}{2} + \phi\right] e^{-q_2\left[y - \left(\frac{G}{2} + W\right)\right]},$$

$$\text{for} \quad \frac{G}{2} + W \leq y \tag{2.31e}$$

where B and B' are related by

$$B'\left[e^{-q_2\frac{G}{2}} + e^{+q_2\frac{G}{2}}\right] = B \cos\left[h_1 \frac{W}{2} - \phi\right]. \tag{2.32}$$

Note that ϕ, q_2 and h_1 of the symmetric mode are obtained from the following transcendental equations derived from the boundary conditions at $y = \pm |G/2|$ and at $y = \pm |W + (G/2)|$:

$$\frac{h_1}{n_1^2} \sin\left(h_1 \frac{W}{2} + \phi\right) = \frac{q_2}{n_2^2} \cos\left(h_1 \frac{W}{2} + \phi\right), \tag{2.33a}$$

$$B\frac{h_1}{n_1^2} \sin\left(h_1 \frac{W}{2} - \phi\right) = B'\frac{q_2}{n_2^2}\left[e^{+q_2\frac{G}{2}} - e^{-q_2\frac{G}{2}}\right], \tag{2.33b}$$

$$h_1^2 + q_2^2 = \left(n_1^2 - n_2^2\right)k^2. \tag{2.33c}$$

(2) Anti-symmetric mode:

$$H_x = -B \cos\left[h_1\left(\frac{W}{2}\right) + \phi\right] e^{+q_2\left[y + \left(\frac{G}{2} + W\right)\right]}$$

$$\beta^2 - q_2^2 = n_2^2, \quad \text{for} \quad y \leq -\frac{G}{2} - W \tag{2.34a}$$

$$H_x = -B \cos\left[h_1\left(y + \frac{G+W}{2}\right) - \phi\right],$$

$$\text{for} \quad -\left(\frac{G}{2} + W\right) \leq y \leq -\frac{G}{2} \tag{2.34b}$$

$$H_x = B'[e^{+q_2 y} - e^{-q_2 y}]$$

$$\beta^2 + h_1^2 = n_1^2 \quad \text{for} \quad -\frac{G}{2} \leq y \leq +\frac{G}{2} \tag{2.34c}$$

$$H_x = B\cos\left[h_1\left(y - \frac{G+W}{2}\right) + \phi\right],$$

$$\text{for} \quad \frac{G}{2} \le y \le \frac{G}{2} + W \tag{2.34d}$$

$$H_x = B\cos\left[h_1\frac{W}{2} + \phi\right]e^{-q_2\left[y - \left(\frac{G}{2}+W\right)\right]},$$

$$\text{for} \quad \frac{G}{2} + W \le y \tag{2.34e}$$

where B and B' are related by

$$B'\left[e^{-q_2\frac{G}{2}} - e^{+q_2\frac{G}{2}}\right] = -B\cos\left[h_1\frac{W}{2} - \phi\right]. \tag{2.35}$$

Note that ϕ, h_1 and q_2 of the anti-symmetric mode are solutions of the following transcendental equations obtained from the boundary conditions:

$$\frac{h_1}{n_1^2}\sin\left(\frac{h_1 W}{2} + \phi\right) = \frac{q_2}{n_2^2}\cos\left(\frac{h_1 W}{2} + \phi\right), \tag{2.36a}$$

$$B'\frac{q_2}{n_2^2}\left[e^{-q_2\frac{G}{2}} + e^{+q_2\frac{G}{2}}\right] = B\frac{h_1}{n_1^2}\sin\left(h_1\frac{W}{2} - \phi\right), \tag{2.36b}$$

$$h_1^2 + q_2^2 = \left(n_1^2 - n_2^2\right)k^2. \tag{2.36c}$$

2.3.1.4   Super modes viewed from coupled mode analysis

It is interesting to note that when the separation between the two waveguides is sufficiently large, coupled mode equations in Eq. (2.11) could be used to show that the super modes of the total structure are just linear combinations of the modes of the uncoupled waveguides.

Again, let the two waveguides in Fig. 2.2(b) be identical. This is the classical example of a pair of coupled identical waveguides. Mathematically, in terms of Eq. (2.11), we have $\Delta\beta = 0$, $\varepsilon_A = \varepsilon_B$, and $C_{AB} = C_{BA} = C$. Then, the solution of Eq. (2.11) is

$$a_A(z) = \frac{1}{2}(A - B)e^{+jCz} + \frac{1}{2}(A + B)e^{-jCz},$$

$$a_B(z) = \frac{1}{2}(B - A)e^{+jCz} + \frac{1}{2}(A + B)e^{-jCz},$$

$$C = \frac{\omega}{4}\iint\limits_{S_B}(\varepsilon_A - \varepsilon_3)[\underline{e}_B \cdot \underline{e}_A]dS. \tag{2.37}$$

Values of A and B were determined from the initial condition at $z = 0$. Substituting this result into Eq. (2.10), we obtain

$$\underline{E}' = \frac{1}{\sqrt{2}}(A-B)\left[\frac{1}{\sqrt{2}}(\underline{e}_A - \underline{e}_B)\right]e^{-j(\beta - C)z} + \frac{1}{\sqrt{2}}(A+B)\left[\frac{1}{\sqrt{2}}(\underline{e}_A + \underline{e}_B)\right]e^{-j(\beta+C)z}. \qquad (2.38)$$

Therefore, any electric field of two identical waveguides can be considered as a superposition of two super modes. The mode which consists of the symmetric combination, $\underline{e}_s = \frac{1}{\sqrt{2}}(\underline{e}_A + \underline{e}_B)$, is a normalized symmetric eigenmode with $\beta_s = \beta + C$. The mode which consists of the anti-symmetric combination, $\underline{e}_a = \frac{1}{\sqrt{2}}(\underline{e}_A - \underline{e}_B)$, is an anti-symmetric eigenmode with $\beta_a = \beta - C$. In other words,[4]

$$\begin{aligned}\underline{E}' &= \frac{1}{\sqrt{2}}(A-B)\underline{e}_a e^{-j\beta_a z} + \frac{1}{\sqrt{2}}(A+B)\underline{e}_s e^{-j\beta_s z}\\ &= A_a \underline{e}_a e^{-j\beta_a z} + A_s \underline{e}_s e^{-j\beta_s z}.\end{aligned} \qquad (2.39)$$

However, the $\underline{e}_s$ and $\underline{e}_a$ used in the coupled mode analysis are just the superposition of the unperturbed $\underline{e}_A$ and $\underline{e}_B$. The modes, β_a and β_s, are wrong when the separation of the two waveguides is small or when the perturbation is strong. In that case there will still be symmetric and anti-symmetric super modes. The correct solution of the modes, β_a and β_s, is given by the super mode analysis.

2.3.1.5 Propagation of super modes in two coupled waveguides with variable gap

When two parallel waveguides have a variable gap between them, the total structure can be approximated by a series of local coupled waveguides connected in series. Within each local section j, there are two parallel waveguides with a constant separation G_j. When super modes are excited at the front end, they propagate from one local section to another in cascade. If the discontinuity of the gap between two adjacent sections is small, there will not be any significant change of the amplitude and phase of the super modes at each junction. This is known as the adiabatic propagation of super modes. Let the length of the jth local section be l_j. Let there be two super modes with complex amplitude (including phase), A_{1j} and A_{2j} at the input end and A'_{1j} and A'_{2j} at the output end. Let the propagation wave numbers, i.e. $n_{\text{eff}}k$s, of the super modes (corresponding to the symmetric and anti-symmetric modes of the case when waveguides are identical) in the jth section be β_1 and β_2. Then the complex amplitudes are related by

$$\begin{Vmatrix} A'_{1j} \\ A'_{2j} \end{Vmatrix} = \begin{Vmatrix} e^{-j\beta_1 l_j} & 0 \\ 0 & e^{-j\beta_2 l_j} \end{Vmatrix} \bullet \begin{Vmatrix} A_{1j} \\ A_{2j} \end{Vmatrix} = \|t_j\| \bullet \begin{Vmatrix} A_{1j} \\ A_{2j} \end{Vmatrix}, \qquad (2.40)$$

and $A'_{1j} = A_{1(j+1)}$ and $A'_{2j} = A_{2(j+1)}$. Therefore,

$$\begin{aligned}\begin{Vmatrix} A_{1,\text{out}} \\ A_{2,\text{out}} \end{Vmatrix} &= \|t_N\| \bullet \|t_{N-1}\| \bullet \bullet \|t_j\| \bullet \bullet \|t_2\| \bullet \|t_1\| \bullet \begin{Vmatrix} A_{1,\text{in}} \\ A_{2,\text{in}} \end{Vmatrix}\\ &= \begin{Vmatrix} e^{-j\zeta_1} & 0 \\ 0 & e^{-j\zeta_2} \end{Vmatrix} \bullet \begin{Vmatrix} A_{1,\text{in}} \\ A_{2,\text{in}} \end{Vmatrix},\end{aligned} \qquad (2.41)$$

where, $\zeta_1 = \sum\limits_{j=1}^{j=N} \beta_{1j}l_j$ and $\zeta_2 = \sum\limits_{j=1}^{j=N} \beta_{2j}l_j$. Here N is the total number of local sections. $A_{1,\text{in}}$ and $A_{2,\text{in}}$ are the input complex amplitudes to the 1st and 2nd super mode, and $A_{1,\text{out}}$ and $A_{2,\text{out}}$ are the output complex amplitudes.

2.3.2   Directional coupling, viewed as propagation of super modes

The symmetric mode, the anti-symmetric mode, and the modes of individual isolated waveguides are illustrated in Fig. 2.4(b). The symmetric mode e_s is the lowest order super mode of the entire structure with the highest effective index. The actual field at any position z (e.g. $z=0$) in the coupled waveguide depends on A and B. When $A=B$, only the symmetric mode exists. When $A=-B$, only the anti-symmetric mode exists. When $B=0$ (or $A=0$), both the symmetric and the anti-symmetric modes exist with equal amplitude. Since the symmetric and the anti-symmetric modes do not have the same phase velocity, the relative phase between the two modes will oscillate as a function of distance of propagation. Consequently the intensity of the total field in waveguides A and B will be a function of z for $0 < z < W$. Let $A = 1$ and $B = 0$ at $z = 0$. When $CW = \pi/2$ in Eq. (2.38), $A = 0$ at $z = W$. We would have transferred all the power from A at $z = 0$ to B at $z = W$. For $z > W$, the two waveguides are well separated from each other where $C = 0$ as shown in Fig. 2.4(a). The symmetric and anti-symmetric modes for $z < 0$ and $z > W$ have the same β as the modes of the individual waveguides. The power in waveguide A and B in those regions is independent of z.

2.3.3   Super modes of two coupled waveguides in general

The discussion in the previous section applies to any two parallel identical waveguides. It can be generalized directly to two non-identical waveguides. In this case the wave-guides A and B in Fig. 2.5(a) will have different indices, widths and thicknesses. The analysis of planar waveguide modes in the x direction applies separately to the cladding and the waveguides. The top view of the waveguides in Fig. 2.5(b) will have different n_{eA} and n_{eB} as well as different widths for waveguides A and B. The effective index approximation can be used again to calculate all the modes in the y direction.[5] However, the first two modes will no longer be symmetric or anti-symmetric, they will be just the zeroth and first order modes. There may even be higher order modes. Note that for TM modes in the x direction, TE effective index analysis will be used in the y direction.

2.3.4   Adiabatic branching and the super mode analysis of the Mach–Zehnder interferometer

In the following subsections, a new concept, the adiabatic transition is introduced first.

2.3.4.1   The adiabatic transition

Consider the transition for a guided-wave mode propagating from waveguide C into waveguide D as shown in Fig. 2.6(a), also known commonly as a waveguide horn. Let waveguide C be a single mode waveguide and waveguide D be a multimode waveguide. As the waveguide cross-section expands, the second mode emerges at $z = z_1$ (i.e. there exists a second mode in the electromagnetic solution of an infinitely long waveguide that has the same transverse dielectric index variation as the cross-sectional index

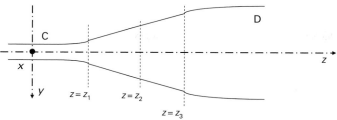

(a) The transition from a single mode channel waveguide to
    a multimode channel waveguide

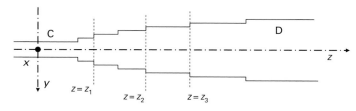

(b) The step approximation of the transition

Fig. 2.6. Top view of an adiabatic transition and its step approximation. (a) The transition from a single mode channel waveguide to a multimode channel waveguide (i.e. a waveguide horn). (b) The step approximation of the transition. Within each local section of the waveguide the dielectric constant profile is independent of z. The 2nd mode exists for $z > z_1$, the 3rd mode exists for $z > z_2$, the 4th mode exists for $z > z_3$.

variation at $z = z_1$). The third mode emerges at $z = z_2$, etc. The transition section can be approximated by many steps of local waveguides that have constant cross-section within each step as shown in Fig. 2.6(b). At each junction of two adjacent steps, modal analysis can be used to calculate the excitation of the modes in the new step by the modes in the previous step. For adiabatic transition in the forward direction, the steps are so small that only the specific order mode is excited in the next section by the same order mode in the previous section. In terms of Eq. (1.51), the overlap integral of the lowest order mode in the transmitted section to the lowest order mode in the incident section is approximately one, while the overlap integrals of the higher order modes in the transmitted section to the lowest order mode in the incident section are approximately 0. In other words, a negligible amount of power is coupled into higher order modes and radiation modes. Therefore, in a truly adiabatic transition, only the lowest order mode is excited in the multimode output waveguide by the lowest order mode in the input section, and there is no power loss. Conversion of power into higher order modes will occur when the tapering is not sufficiently adiabatic or when there is scattering. The same conclusion can be drawn for propagation of the lowest order mode in the reverse direction, i.e. from D to C.

Let us now consider a reverse transition from $z > z_3$ to $z = 0$ where the incident field excited several modes at D. Whenever a higher order mode propagating in the $-z$ direction is excited at D, it will not be transmitted to C. The power in this higher order mode will be transferred into the radiation modes at the z position where this mode is cut-off. Only the power in the lowest order mode at D will be transmitted to the lowest order mode at C. *An important practical significance of this result is that when a LED is used to excite a single mode waveguide via a waveguide horn, the excitation efficiency will be very low.*

2.3.4.2   Super mode analysis of a symmetric Y-branch
(A) Y-branch of a single mode waveguide
A guided-wave component used frequently in fiber and channel waveguide devices is a symmetric Y-branch. Its plan view in the yz plane is illustrated in Fig. 2.7(a). The single mode channel waveguide at $z = 0$ is connected to two single mode channel waveguides. The Y-branch is symmetric in the y direction with respect to the xz plane. The waveguides at $z > L_0$ have large separation distance and identical cross-sectional index profile in the y direction. The index profile in the x direction is uniform for the entire device. In other words, a single mode waveguide at $z = 0$ is split into two uncoupled identical single mode waveguides at $z > L_0$. The practical application of such a device is to split the forward

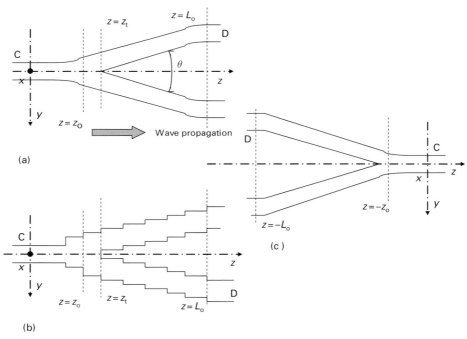

Fig. 2.7.   Top view of a symmetric Y-branch coupler. (a) A symmetric 3 dB coupler that splits the power in the input single mode channel waveguide at C equally into two identical channel waveguides at D. For a symmetric Y-branch, the modes of individual waveguides at D always have equal amplitude and phase. (b) The step approximation of the Y-branch 3 dB coupler. (c) The reverse symmetric coupler that combines the fields from two input waveguides into a single mode output waveguide at C. The power in the symmetric super mode at D will be transmitted without loss into the mode of the single mode waveguide at C. Power in the anti-symmetric super mode will be radiated.

propagating power in the original waveguide at C equally into two waveguides at D where they are well separated from each other. It is an adiabatic transition when the angle of the branching, θ, is sufficiently small such that the scattering and conversion loss from $z = 0$ to $z = L_0$ can be neglected. Ideally, a symmetric Y-branch should function like a 3dB coupler from the input to both outputs.

The forward propagating Y-branch coupler can be analyzed as follows. In Fig. 2.7(a), the input waveguide has a single TE_0 mode at $z < 0$. The waveguide width in the y direction begins to broaden at $z > 0$. At $z > z_0$, the waveguide (or the split waveguides) has two modes. At $z > z_t$, there are two waveguides. From $z = z_t$ to $z \cong L_0$, since each isolated waveguide has a single TE_0 mode, e_A and e_B, the two super modes, are the symmetric mode and the anti-symmetric mode, discussed in Sections 2.3.1.3 and 2.3.1.4. From the symmetry point of view, no anti-symmetric mode is excited in an adiabatic transition. At $z > L_0$, the coupling between two waveguides is zero, thus the optical power is split equally into waveguides A and B.

In the reverse situation shown in Fig. 2.7(c), when the incident field is the lowest order symmetric mode of the double waveguides it is transmitted without loss to the output waveguide as the TE_0 mode. However, if the incident mode is an anti-symmetric mode, it will continue to propagate as the anti-symmetric mode from $z = -L_0$ to its cut-off point. Let the cut-off point be $z = -z_0$. At z just before $-z_0$, the anti-symmetric mode will be very close to cut-off, with a very long evanescent tail, and its n_{eff} is very close to the effective index of cladding or substrate modes. As z approaches $-z_0$, the anti-symmetric mode begins to transfer its energy into the radiation mode in the cladding or the substrate. Because of the small overlap integral between the anti-symmetric and the TE_0 mode, the TE_0 mode will not be excited by the anti-symmetric mode. Similar comments can be made for any other higher order mode excited at $z < -L_0$. It will be coupled to radiation modes at its cut-off point. *In summary, only the power in the lowest order symmetric mode will be transferred to the* TE_0 *mode at the output.*

(B) Y-branch of a double mode waveguide and the Y-branch reflector

The preceding conclusion will not necessarily apply if the waveguide at $z = 0$ in Fig. 2.7(a) has two modes. Let the two modes at $z = 0$ be symmetric and anti-symmetric modes. In the case of a forward propagating Y-branch, if the incident radiation is just in the symmetric mode, it will be transmitted as the symmetric mode at $z = L_0$ as discussed in the preceding paragraph. If the incident radiation is just in the anti-symmetric mode, it will be transmitted as the anti-symmetric mode at $z = L_0$. In either case, the incident power will still be split equally into the two waveguides, A and B. However, for the anti-symmetric excitation, there will be a π phase difference between the amplitudes of the modes in waveguides A and B. Symmetric and anti-symmetric modes have different phase velocities, i.e. n_{eff}. When both the symmetric and anti-symmetric mode are excited at $z = 0$ the resultant field at $z = L_0$ will be very different depending on the phase velocity difference and the distance of propagation. For equal amplitude of symmetric and anti-symmetric mode at $z = 0$, the power split could vary from 100% in waveguide A and zero in waveguide B to 100% in waveguide B and zero in waveguide A. The split depends on the index profiles, the branching angle and the distance of propagation.

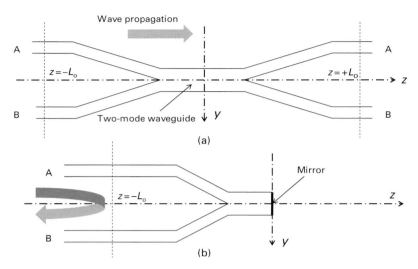

Fig. 2.8.   The Y-branch power splitter and reflector. (a) Two symmetric Y-branch couplers connected back-to-back with a two-mode connecting waveguide. The ratio of the output power transmitted from the input into waveguides A and B will depend on the relative phase of the symmetric and anti-symmetric modes at $z = +L_0$. (b) The symmetric Y-branch reflector with a two-mode waveguide. The ratio of the reflected power in waveguides A and B will depend on the relative phase of the reflected symmetric and anti-symmetric modes at $z = -L_0$.

In the reverse coupler shown in Fig. 2.7(c), when the waveguide at $z = 0$ has two modes, radiation in both the symmetric and anti-symmetric mode will be transmitted without loss to $z = 0$. However, the total field pattern at $z = 0$ will be very different depending on the relative phase between them, which is the total cumulative phase difference between the two modes from $z = -L_0$ to $z = 0$.

If there are two Y-branch couplers connected back-to-back as shown in Fig. 2.8(a) where the single waveguide is a two-mode waveguide, and if the input power at $z = -L_0$ is all in waveguide A, then equal amplitudes of symmetric and anti-symmetric mode are excited at $z = -L_0$. The output at $z = L$ could then vary from 100% power in waveguide A to 100% power in waveguide B, or any other split ratio, depending on the cumulative phase difference between the symmetric and anti-symmetric mode from $z = -L_0$ to $z = L_0$. Figure 2.8(b) shows a Y-branch coupler with the single two-mode waveguide terminated at a mirror. It is simply a back-to-back Y-branch as shown in Fig. 2.8(a) folded over. Thus its analysis is identical to the previous analysis. However, it has a new practical significance. It implies that, for power incident into waveguide A, we can make a Y-branch reflector with a controlled fraction of power reflected back into waveguides A and B. *Note that, from the point of view of super mode analysis, the two-mode waveguide is just a directional coupler with zero gap of separation.*

2.3.4.3   Wave propagation in an asymmetric Y-branch

Note that the analysis based on symmetric and anti-symmetric modes applies only to symmetrical adiabatic Y-branches. When the branching angle is large in non-adiabatic transitions, mode conversion will occur at step junctions. When the branches are not symmetrical, the local super modes could have very asymmetrical electromagnetic field profiles. Conversion between super modes might occur at each step junction. The output, i.e. the cumulative effect, will depend on initial conditions, the branching angle, the index profile and the asymmetry of the Y-branch. An asymmetrical Y-branch will behave sometimes as a power divider and sometimes as a mode splitter or converter [9]. Numerical analysis based on modal analysis of the super modes in the step approximation is required to find the answer.

2.3.4.4   The Mach–Zehnder interferometer

The Mach–Zehnder interferometer consists of two symmetric Y-branches connected by two parallel channel waveguides which are well separated from each other so that they are uncoupled. It is illustrated in Fig. 2.9. Similar devices can be made from optical fibers. The objective of the input forward Y-branch in the Mach–Zehnder interferometer is to excite equally the individual modes of the two waveguides, i.e. the symmetric super mode, immediately after the input Y-branch. The object of the connecting waveguides is to provide a specific phase difference between the two super modes at the entrance of the output reverse Y-branch coupler. Only the symmetric mode excited at the reverse output coupler is transmitted as the output, while the anti-symmetric mode is radiated away as the substrate radiation modes.

Let the input be a TE_0 mode with amplitude A at $z = 0$. At the exit of the input Y-branch at $z = L_b$, the amplitude of the symmetric mode is $A \exp(j\phi)$. Note that ϕ is the phase shift due to the propagation from $z = 0$ to $z = L_b$. In terms of the modes of the individual waveguides, the amplitudes are $\frac{1}{\sqrt{2}} A e^{j\phi} \underline{e}_A$ and $\frac{1}{\sqrt{2}} A e^{j\phi} \underline{e}_B$. The modes of the two uncoupled connecting waveguides are $\underline{e}_A$ and $\underline{e}_B$. When the two parallel

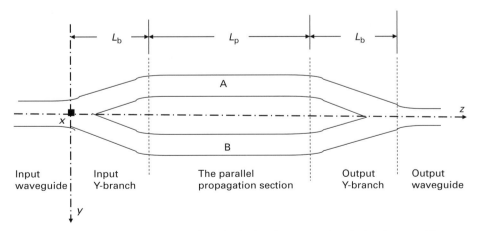

Fig. 2.9.   Top view of a channel waveguide Mach–Zehnder interferometer. Two isolated waveguides, A and B, connect an input symmetrical Y-branch 3 dB coupler to an output reverse symmetrical Y-branch coupler. Waveguides A and B are well separated from each other. Only the power in the symmetric mode at the input of the output Y-branch coupler will be transmitted to the single mode output waveguide.

waveguides are identical and have equal length L_p, the input to the output Y-branch at $z = L_b + L_p$ is $\frac{1}{\sqrt{2}} A e^{j\phi} e^{-j\beta_A L_p} (\underline{e}_A + \underline{e}_B)$. Such a symmetric mode will yield an output $A e^{j2\phi} e^{-j\beta_A L_p}$ at $z = 2L_b + L_p$.

When the two parallel waveguides in the propagation section have slightly different effective index or propagation wave number, β_A and β_B, the input to the output Y-branch is

$$\frac{1}{\sqrt{2}} A e^{j\phi} e^{-j\beta_A L_p} \left(\underline{e}_A + \underline{e}_B e^{-j(\beta_B - \beta_A)L_p} \right). \tag{2.42}$$

In other words, there is a mixture of symmetric mode, $\underline{e}_s$, and anti-symmetric mode, $\underline{e}_a$, at $z = L_b + L_p$. When $\Delta\beta L_p = (\beta_B - \beta_A)L_p = \pm\pi$ or $(2n \pm 1)\pi$, where n is an integer, then the input to the output Y-branch is an anti-symmetric mode. In this case, the output TE_0 mode at $z = 2L_b + L_p$ will have zero amplitude. The power in the anti-symmetric mode is transferred into the radiation modes. When the power transmitted to the output is calculated based on the amplitude of the symmetric mode at $z = L_b + L_p$, we obtain,

$$\frac{I_{\text{out}}}{I_{\text{in}}} = \frac{1}{2} \left[1 + \cos(\Delta\beta L_p) \right]. \tag{2.43}$$

Such a device is called a Mach–Zehnder interferometer.

The super mode analysis is important to understand the Mach–Zehnder modulator in depth. For example, when the attenuation of one of the waveguides is very large, e.g. the B waveguide, then the input to the output Y-branch is $\frac{1}{\sqrt{2}} A e^{j\phi} e^{-j\beta_A L_p} \underline{e}_A = \frac{A}{2} (\underline{e}_s + \underline{e}_a) e^{j\phi} e^{-j\beta_A L_p}$. Since only $\underline{e}_s$ will be transmitted, the amplitude of the TE_0 mode at the output is $\frac{A}{2} e^{2j\phi} e^{-j\beta_A L_p}$. In other words, only 1/4 of the input power is transmitted, 3/4 of the input power is attenuated and radiated into the cladding or the substrate.

2.4   Propagation in multimode waveguides and multimode interference couplers

Interference of modes in a multimode waveguide has interesting and important applications. A multimode interference coupler consists of a section of multimode channel waveguide, abruptly terminated at both ends. A number of access channel waveguides (usually single mode) may be connected to it at the beginning and at the end. Such devices are generally referred to as NxM multimode interference (MMI) couplers where N and M are the number of input and output waveguides respectively [10].

Figure 2.10(a) illustrates a multimode interference coupler with two input and two output access waveguides. The multimode section is shown here as a step-index ridge waveguide with width W and length L. It is single mode in the depth direction x and multimode ($n \geq 3$) in the lateral direction y. The objective of such a multimode coupler is to couple specific amounts of power from the input access waveguides into the output access waveguides. Its operation is based on the interference of the propagating modes.

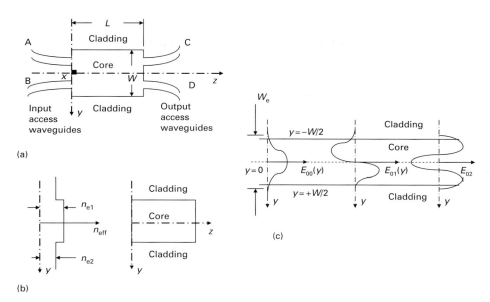

Fig. 2.10. A multimode interference coupler. (a) The top view of a 2×2 multimode interference coupler. The multimode waveguide is L long and W wide. (b) The effective index profile of the multimode waveguide. (c) The field patterns (as a function of y) of the lowest order modes of the multimode section.

Based on the interference pattern we will show that various distributions of the power in the output access waveguides can be obtained at different z positions.

Let the multimode waveguide be a ridge waveguide as shown in Fig. 1.7(b). For the planar waveguide mode (i.e. for very large W) in the core (i.e. in the ridge), it has just a single TE mode in the x direction with an effective index n_{e1}. The cladding region, outside the ridge, also has a planar waveguide mode with an effective index n_{e2}. $n_{e1} > n_{e2}$. Figure 2.10(b) illustrates the profile of the effective index of the planar TE_0 modes in the y direction. The channel guided-wave modes in the core can be found by the effective index method discussed in Section 1.2.6 or by other numerical methods. Figure 2.10(c) illustrates the effective mode width, W_e, and the lateral field variation in the y direction for the first few modes.

Before we discuss the interference pattern of the modes, let us consider first the properties of individual modes. For well guided modes, it has been shown in the literature [11] that the solution of transcendental Eq. (1.46) can be approximated by

$$\tan\left[(h_n/h_n kk)\frac{kW_e}{2}\right] \cong \infty.$$

Here W_e is an effective width of the ridge, $W_e > W$. Note that W_e is usually taken to be the effective width of the lowest order mode $m = 0$ in the x direction and $n = 0$ in the y direction. In that case

$$h_n = \frac{(n+1)\pi}{W_e},$$

and

$$\beta_{0n}^2 = n_{e1}^2 k^2 - h_n^2, \qquad \beta_{0n} \cong n_{e1}k - \frac{(n+1)^2 \pi \lambda}{4n_{e1}W_e^2}. \tag{2.44}$$

Equation (2.44) predicts that the propagation constants of the various lateral order modes will have a quadratic dependence on n. By defining L_π as the beat length (i.e. the propagation length in which the phase difference of two modes is π) between the $n = 0$ and $n = 1$ modes, we obtain

$$L_\pi = \frac{\pi}{\beta_{00} - \beta_{01}}, \qquad \beta_{00} - \beta_{0n} = \frac{n(n+2)\pi}{3L_\pi}. \tag{2.45}$$

Let us now examine the total field of all the modes. As we have discussed in Section 1.2.7, the y variation of any input field at $z = 0$, $E_0(y,z=0)$, can be expressed as a summation of the E_{0n} modes. Thus

$$E_0(y, 0) = \sum_{n=0}^{n=N-1} C_n E_{0n}(y),$$

$$E_0(y, z) = \sum_{n=0}^{n=N-1} \left\{ C_n E_{0n}(y) e^{\left[j \frac{n(n+2)\pi}{3L_\pi} z \right]} \right\} e^{-j\beta_{00}z},$$

$$E_{0n}(y) = A \sin(h_n y). \tag{2.46}$$

Any input field at $z = 0$ will be repeated or mirrored at $z = L$, whenever

$$\exp \left[j \frac{n(n+2)\pi}{3L_\pi} L \right] = 1, \tag{2.47}$$

or

$$\exp \left[j \frac{n(n+2)\pi}{3L_\pi} L \right] = (-1)^n. \tag{2.48}$$

When the condition in Eq. (2.47) is satisfied, the field at $z = L$ is a direct replica of the input field. When the condition in Eq. (2.48) is satisfied, the even modes will have the same phase as the input, but the odd modes will have a negative phase, producing a mirrored image of the input field. For the 2×2 coupler shown in Fig. 2.10(a), it means that power in input A will be transferred to output C when Eq. (2.47) is satisfied. Power in input A will be transferred to output D when Eq. (2.48) is satisfied.[6]

More extensive use of the mode interference pattern can be obtained when we analyze it in detail in the following manner. Figure 2.10(c) shows that the y variation of the field of a well guided multimode channel waveguide mode resembles the lowest order sine terms of a Fourier series in y within the period from $y = -W_e/2$ to $y = +W_e/2$. However, there is only a finite number of sine Fourier series terms in our modes. In order to recognize the more complex interference patterns, let us now extend the expression for the modes to outside of the range $-W_e/2$ to $W_e/2$ in a periodic manner so that we can take advantage of our knowledge of Fourier series. Since these modes have a half-cycle sine variation within $-W_e/2 < y < W_e/2$, the extended mode in $-3W_e/2 < y < -W_e/2$ and in $W_e/2 < y < 3W_e/2$ should be anti-symmetric with respect to the mode in $-W_e/2 < y < W_e/2$. Similar extension can be made beyond $y > |3W_e/2|$. Consider now the total extended field over all y coordinates, including the periodic extension of the fields outside the multimode waveguide region. The extended input field from all the input access waveguides (periodically repeated outside the region from $y = -W_e/2$ to $W_e/2$) could then be expressed as a summation of these Fourier terms. Equation (2.46) shows that at a distance L the relative phase among the Fourier terms is changed. Different multi-fold images can be formed within the period, ranged from $-W_e/2$ to $W_e/2$, by manipulating these phase terms. As an example, let us consider $L = 3pL_\pi/2$ where p is an odd integer. Then

$$E_0\left(y, \frac{3pL_\pi}{2}\right) = \sum_{n \text{ even}} C_n E_{0n}(y) + \sum_{n \text{ odd}} (-j)^p C_n E_{0n}(y)$$

$$= \frac{1 + (-j)^p}{2} E_0(y, 0) + \frac{1 - (-j)^p}{2} E_0(-y, 0). \quad (2.49)$$

The last equation represents a pair of images of E_0 in quadrature and with amplitudes $1/\sqrt{2}$, at distances $z = 3L_\pi/2$, $9L_\pi/2$ .... The replicated, the mirrored, and the double images of E_0 at various z distances are illustrated in Fig. 2.11. Clearly, we have a 3 dB power splitter from the input B into output waveguides, C and D, at $z = 3L_\pi/2$ and at $z = 9L_\pi/2$. We have transferred the power from B to C (called the cross-state) when $z = 3L_\pi$, and from B to D (called the through-state) when $z = 6L_\pi$. Note that a 2×2 InGaAsP MMI cross coupler has been made with $W = 8\,\mu m$ and $L = 500\,\mu m$ which gives an excess loss of 0.4 to 0.7 dB and extinction ratio of 28 dB, and a 3 dB splitter with $L = 250\,\mu m$ and imbalances between C and D well below 0.1 dB [10].

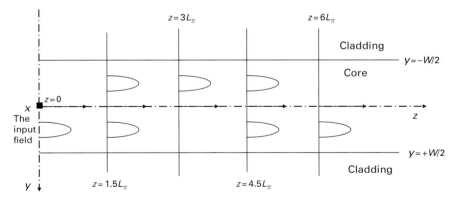

Fig. 2.11. Images of the input field at various distances in a multimode interference coupler. The input field is shown at $z = 0$. It can be decomposed into a summation of all the modes. Each mode has a different phase velocity. The total field profile of the summation of these modes will yield a two-fold image of the input at $z = 1.5L_\pi$ and at $z = 4.5L_\pi$, a mirror single image at $z = 3L_\pi$, and a direct single image at $z = 6L_\pi$.

The actual design of an MMI coupler must take into account the number of input and output access waveguides, the number of modes in the multimode waveguide, the relative phase and amplitude of the incident modes in the input access waveguides and the position and width of access waveguides.

Notes

1. The orthogonality relation applies only to modes in lossless waveguides.
2. The orthogonality conditions are not applicable to the C_{ij} integrals in Eq. (2.11) if ε_is are spatially variant in the integral. Orthogonality has been proven only for lossless waveguides.
3. Instead of varying $\Delta\beta$, C could also be varied in order to change I_{out}/I_{in}.
4. It has also been shown by coupled mode analysis [2] that when waveguides A and B are not identical, there are still two super modes, the propagation wave numbers of which are:

$$\beta = \frac{\beta_A + \beta_B}{2} \pm \sqrt{C_{BA} C_{AB} + \sqrt{(\Delta\beta/2)^2}}.$$

5. There may be more than one planar waveguide mode in the x direction. In that case effective index analysis in the y direction must be used separately for each mode in the x direction. Note that the effective index approximation is effective only when modes with similar $E_m(x)$ in the x direction are used in the calculation of the y variation.
6. Note that a two-mode interference coupler is identical to a two-waveguide directional coupler with zero gap of separation. Therefore it can also be analyzed by super mode analysis. This concept can also be extended to an MMI coupler.

References

1. A. Yariv, *Optical Electronics in Modern Communication*, Oxford University Press (1997).
2. W. S. C. Chang, *Principles of Lasers and Optics*, Cambridge University Press (2005).
3. R. Ulrich and P. K. Tien, Theory of prism-film coupler and guide. *J. Opt. Soc. Am.*, **60** (1970) 1325.
4. D. L. Lee, *Electromagnetic Principles of Integrated Optics*, Chapter 8, John Wiley and Sons (1986).
5. A. Yariv, Coupled mode theory for guided wave optics. *IEEE J. Quant. Electr.*, **QE-9** (1973) 919.
6. D. C. Flanders *et al.*, Grating filters for thin film optical waveguides. *Appl. Phys. Lett.*, **24** (1974) 195.
7. S. Kurazono, K. Iwasaki, and N. Kumagai, New optical modulator consisting of coupled optical waveguides. *Elect. Comm. Japan*, **55** (1972) 103.
8. H. Kogelnik and R. V. Schmidt, Switched directional couplers with alternating $\Delta\beta$. *IEEE J. Quant. Elect.*, **QE-12** (1976) 396.
9. W. K. Burns and A. F. Milton, Mode conversion in planar dielectric separating waveguides. *IEEE J. Quant. Electr.*, **QE-11** (1975) 32.
10. L. B. Soldano and E. C. M. Pennings, Optical multi-mode interference devices based on self-imaging: Principles and applications. *J. Lightwave Tech.*, **13** (1995) 615.
11. N. S. Kapany and J. J. Burke, *Optical Waveguides*, Academic Press (1972).

3   Electro-optical effects

Electro-optical effects in materials are used to switch, modulate, detect, amplify, or generate optical radiation in guided-wave devices.

The best known electro-optical effect is probably the amplification of optical radiation by stimulated emission of radiation. In edge-emitting semiconductor lasers, the amplification of the guided wave is obtained via current injection in a forward biased p–n junction. In a laser oscillator, the waveguide is terminated by reflectors (or coupled to a feedback grating) to form a resonant cavity. When amplification exceeds losses in the cavity, oscillation is obtained [1, 2]. When end reflections (or feedback) are absent and when there is net gain, a laser amplifier is obtained [3, 4]. The second well known electro-optical effect is detection of optical radiation by photo-generation of carriers. For optical radiation incident on a semiconductor with photon energy greater than the semiconductor bandgap, electrical carriers are generated by the absorption of incident radiation. In a semiconductor detector, photo-generated carriers in a reverse biased p–i–n junction are collected and transmitted to the external circuit [5]. In waveguide photo-detectors, the optical radiation is incident on to and absorbed in a waveguide so that the absorption can be distributed over a distance, enabling the detector to absorb more effectively the incident optical power over a longer distance while maintaining a large operation bandwidth [6]. Discussion of carrier injection, stimulated emission and carrier transport in semiconductor junctions requires extensive review of semiconductor device physics. There are already many books on lasers and detectors [7–9]. Therefore, guided-wave lasers and detectors will not be discussed here.

Other well known electro-optical effects used in guided-wave devices are the changes of the absorptive or refractive properties of materials created by an applied electrical field or acoustic strain. How they work, and how they are utilized in modulation or switching, will be the focus of the discussion in this book.

Electro-optical effects affect the propagation of guided waves via the susceptibility of the material χ. In general χ is complex,

$$\chi = \chi' - j\chi''. \tag{3.1}$$

In a lossless isotropic material and without any electro-optical effect, we have assumed in Chapters 1 and 2 that

$$\varepsilon = \chi_0 \varepsilon_0 = n^2 \varepsilon_0, \quad \text{or} \quad \chi'_0 = n^2 \quad \text{and} \quad \chi''_0 = 0. \tag{3.2}$$

When there is an electro-optical effect, there is a change of χ from χ_0 to χ_{eo}, by $\Delta\chi$. $\Delta\chi$ has a real part $\Delta\chi'$ and an imaginary part $\Delta\chi''$,

$$\Delta\chi = \Delta\chi' - j\Delta\chi'', \tag{3.3}$$

$$\varepsilon = \chi_{eo}\varepsilon_0 = (\chi_0 + \Delta\chi)\varepsilon_0 = n^2\varepsilon_0 + (\Delta\chi)\varepsilon_0. \tag{3.4}$$

In general, $\chi_{eo} = \chi'_{eo} - j\chi''_{eo}$, thus the real and the imaginary part of χ_{eo} are

$$\chi'_{eo} = n^2 + \Delta\chi' \quad \text{and} \quad \chi''_{eo} = \Delta\chi''. \tag{3.5}$$

For plane waves propagating in the z direction in a material that has susceptibility χ_{eo}

$$E(z, t) = E e^{j(\omega t - k_{eo}z)}. \tag{3.6}$$

Since

$$k_{eo} = \omega\sqrt{\mu_0\varepsilon_{eo}} \cong \omega\sqrt{\mu_0 n^2\varepsilon_0}\left[\left(1 + \frac{\Delta\chi'}{2n^2}\right) - j\left(\frac{\Delta\chi''}{2n^2}\right)\right], \tag{3.7}$$

we obtain

$$E(z, t) = \left[E e^{j\omega t} e^{-j\omega\sqrt{\mu_0 n^2\varepsilon_0}z}\right] e^{-j\omega\sqrt{\mu_0 n^2\varepsilon_0}\frac{\Delta\chi'}{2n^2}z} e^{-\frac{\omega\sqrt{\mu_0 n^2\varepsilon_0}\Delta\chi''}{2n^2}z}, \tag{3.8}$$

and

$$I(z) = \frac{2}{\sqrt{\mu_0/\varepsilon_{eo}}}|E(z, t)|^2 \cong \left[\frac{2E^2}{\sqrt{\mu_0/n^2\varepsilon_0}}\right] e^{-\frac{\omega\sqrt{\mu_0 n^2\varepsilon_0}\Delta\chi''}{n^2}z}. \tag{3.9}$$

We note that

$$\alpha\frac{dI}{dz} = -\alpha I, \quad \text{with} \quad \alpha = \frac{\omega\sqrt{\mu_0\varepsilon_0}\Delta\chi''}{n}. \tag{3.10}$$

Therefore a plane wave exhibits an additional electro-optical phase shift, $\omega\sqrt{\mu_0\varepsilon_0}\frac{\Delta\chi'}{2n}z$, after propagating a distance z. For positive $\Delta\chi''$, the intensity I of the plane wave is attenuated by $\exp\left(-\frac{\omega\sqrt{\mu_0\varepsilon_0}\Delta\chi''}{n}z\right)$. For negative $\Delta\chi''$, the intensity I is amplified.

For guided waves, the effect of electro-optical $\Delta\chi'$ and $\Delta\chi''$ in the material will create a change in the effective index and attenuation of the guided wave discussed in Chapters 1 and 2.

3.1   The linear electro-optic Pockel's effect

The most commonly used electro-optic effect is the linear Pockel's effect, in which the change of the refractive index of the material, or $\Delta\chi'$, is linearly proportional to the applied electric field. Waveguide materials that exhibit a linear electro-optic effect include single crystals such as $LiNbO_3$. Unfortunately, material such as $LiNbO_3$ is anisotropic. Thus the discussion of the electro-optic effect must begin with a discussion of wave propagation in an isotropic medium.

In isotropic materials, the displacement vector $\underline{D}$ and the electric field vector $\underline{E}$ are always parallel to each other, i.e. $\underline{D} = \varepsilon\varepsilon_0\underline{E}$. Note that ε is commonly known as the dielectric constant of the material, and ε_0 is the permittivity of free space. In anisotropic materials, $\underline{E}$ is not generally parallel to $\underline{D}$. The $\underline{D}$ and $\underline{E}$ are related to each other by a dielectric tensor $\underline{\varepsilon}$, $\underline{D} = \varepsilon_0 \underline{\varepsilon} \bullet \underline{E}$, or

$$\begin{vmatrix} D_x \\ D_y \\ D_z \end{vmatrix} = \varepsilon_0 \begin{vmatrix} \varepsilon_x & 0 & 0 \\ 0 & \varepsilon_y & 0 \\ 0 & 0 & \varepsilon_z \end{vmatrix} \begin{vmatrix} E_x \\ E_y \\ E_z \end{vmatrix}. \tag{3.11}$$

Here x, y and z, are known as the principal axes of the material. Only when the electric field is polarized along one of the principal axes will $\underline{D}$ be parallel to $\underline{E}$. The linear electro-optic effect produces a change in $\underline{\varepsilon}$ linearly proportional to an applied electric field $\underline{F}$, thereby affecting the propagation of the guided wave.

A rigorous solution of guided-wave modes using vector Maxwell's equations in an anisotropic medium is very complex. A rigorous analysis of the change of the guided-wave solutions due to a change of $\underline{\varepsilon}$ which is affected by the applied electric field is even more complex. Fortunately, the analysis of a low order TE or TM guided-wave mode can be approximated by the scalar wave equation analysis of the dominant electric or magnetic field discussed in Section 1.2.8. In the scalar wave approximation, the dominant electric and magnetic fields are perpendicular to each other and perpendicular to the direction of propagation of guided waves. The simplest analysis of propagation of a wave with $\underline{D}$ and $\underline{H}$ perpendicular to the direction of propagation and perpendicular to each other in an anisotropic medium is a plane wave analysis. Therefore, we will first review the optical plane wave analysis in an anisotropic medium. We will then show that simple scalar approximate solutions of guided-wave modes that have the dominant electric field polarized in certain specific directions can be obtained quickly from plane wave solutions.

3.1.1 The electro-optic effect in plane waves

From vector Maxwell's equations, optical plane waves in an anisotropic homogeneous lossless material, propagating along a given direction of propagation $\underline{s}$ with propagation wave vector $\underline{\beta}$ ($\underline{\beta} = \beta_x \underline{i_x} + \beta_y \underline{i_y} + \beta_z \underline{i_z} = n\omega\sqrt{\mu\varepsilon_0}(s_x \underline{i_x} + s_y \underline{i_y} + s_z \underline{i_z})$), will have an n value satisfying the equation [10]

$$\frac{s_x^2}{n^2 - \mu\varepsilon_x} + \frac{s_y^2}{n^2 - \mu\varepsilon_y} + \frac{s_z^2}{n^2 - \mu\varepsilon_z} = \frac{1}{n^2}. \tag{3.12}$$

This is a quadratic equation for n. Therefore, for each direction of propagation there are only two plane wave solutions. The $\underline{D}$ for these two plane waves must also satisfy the relationships

$$\frac{D_x^2}{\varepsilon_x} + \frac{D_y^2}{\varepsilon_y} + \frac{D_z^2}{\varepsilon_z} = \underline{E} \bullet \underline{D} \tag{3.13}$$

and

$$\underline{D} \bullet \underline{s} = D_x s_x + D_y s_y + D_z s_z = 0. \tag{3.14}$$

Let $n_x^2 = \varepsilon_x$, $n_y^2 = \varepsilon_y$ and $n_z^2 = \varepsilon_z$. If we define $D_x/\sqrt{\underline{E} \bullet \underline{D}} = x$, $D_y/\sqrt{\underline{E} \bullet \underline{D}} = y$, and $D_z/\sqrt{\underline{E} \bullet \underline{D}} = z$, we obtain

$$\frac{x^2}{n_x^2} + \frac{y^2}{n_y^2} + \frac{z^2}{n_z^2} = 1. \tag{3.15}$$

Equation (3.15) has the mathematical form of an ellipsoid in x, y and z. It is known as the index ellipsoid of the material. Mathematically, it has been shown that the solution of Eq. (3.12), (3.14) and (3.15) can be obtained geometrically as follows. Let us construct first an ellipsoid representing Eq. (3.15) in the x, y and z space. Propagation direction $\underline{s}$ is represented by a vector from the origin in that direction. Let us construct a plane passing through the origin and perpendicular to $\underline{s}$, and the plane intersects the ellipsoid. The intersection is an ellipse, and this ellipse has a major and a minor axis. The length of each axis is the solution of n in Eq. (3.12) and the direction of that axis is the direction of the solution of $\underline{D}$ in Eq. (3.14) and (3.15) [1 (section 9), 10].

Consider LiNbO$_3$ as an example. It is a uniaxial anisotropic crystal with $n_x = n_y = n_o$ and $n_z = n_e$. Note that n_o is known as the ordinary index and n_e is known as the extra-ordinary index. Figure 3.1 shows the index ellipsoid for LiNbO$_3$. It is shown as an ellipsoid with n_o as the axes in x and y and n_e as the axis in z. The cross-section of the ellipsoid in the xy plane is a circle with radius n_o. The $\underline{s}$ is shown as a vector in the xz plane in this example. The intersection of the plane perpendicular to $\underline{s}$ and the index ellipsoid is shown as the shaded ellipse. The major and minor axes are marked as $\underline{D}_e$ and $\underline{D}_o$. For $\underline{s}$ in the xz plane, one of the axes is the y axis. For $\underline{D}$ polarized along the y axis, i.e. $\underline{D}_o$, n will have the value n_o. This is the ordinary wave. The other axis $\underline{D}_e$ is tilted in the xz plane. For $\underline{D}$ polarized along the direction, $\underline{D}_e$, n will have a value between n_o and n_e. The plane wave polarized in this direction is known as an extra-ordinary wave. Since the index ellipsoid is symmetric in x and y, the conclusion about the ordinary wave and extra-ordinary wave applies when $\underline{s}$ is oriented in any direction. Note that when $\underline{s}$ is along the x or y axis, there is a plane wave with $\underline{D}$ polarized along z, this plane wave has $n = n_e$. The second plane wave has $\underline{D}$ polarized along y or x, this plane wave has $n = n_o$. When $\underline{s}$ is polarized along z, both plane waves have $n = n_o$. The plane wave solutions are degenerate, meaning the propagation wave numbers of the two plane waves are identical, no matter what direction the polarization of D in the xy plane.

When a RF[1] or a DC electric field $\underline{F}$ is applied to the electro-optic material, it tilts the index ellipsoid and changes the values of the ellipsoid along its axes. The new ellipsoid will have new axes, x', y' and z'. Mathematically, it means the functional form of the ellipsoid in terms of the crystalline principal axes, x, y, and z, will be changed from that given in Eq. (3.12). The generalized equation for the index ellipsoid is [11]

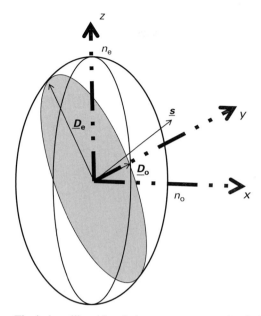

Fig. 3.1. The index ellipsoid and plane wave propagation in LiNbO$_3$. The projection of the ellipsoid on the xy plane is a circle with radius n_o. The projection of the ellipsoid on the yz plane is an ellipse with major axis n_o along y and minor axis n_e along z. The intersection of the plane (perpendicular to $\underline{s}$ and through the origin) with the index ellipsoid is the shaded ellipse. $\underline{D}_o$ is oriented along the major axis of the shaded ellipse, and $\underline{D}_e$ is oriented along the minor axis of this ellipse.

$$\left(\frac{1}{n^2}\right)_1 x^2 + \left(\frac{1}{n^2}\right)_2 y^2 + \left(\frac{1}{n^2}\right)_3 z^2 + 2\left(\frac{1}{n^2}\right)_4 yz + 2\left(\frac{1}{n^2}\right)_5 zx + 2\left(\frac{1}{n^2}\right)_6 xy = 1. \quad (3.16)$$

In the absence of $\underline{F}$,

$$\left(\frac{1}{n^2}\right)_1 = \frac{1}{n_x^2}, \quad \left(\frac{1}{n^2}\right)_2 = \frac{1}{n_y^2}, \quad \left(\frac{1}{n^2}\right)_3 = \frac{1}{n_z^2} \quad \text{and} \quad \left(\frac{1}{n^2}\right)_4 = \left(\frac{1}{n^2}\right)_5 = \left(\frac{1}{n^2}\right)_6 = 0,$$

and Eq. (3.16) is reduced to Eq. (3.15). The change in $(1/n^2)_i$ with $i = 1, 2, 3, 4, 5,$ or 6 is $\Delta (1/n^2)_i$. Note that $\Delta (1/n^2)_i$ are proportional to the applied electric field $\underline{F}^2$ where

$$
\begin{vmatrix}
\Delta\left(\frac{1}{n^2}\right)_1 \\[6pt]
\Delta\left(\frac{1}{n^2}\right)_2 \\[6pt]
\Delta\left(\frac{1}{n^2}\right)_3 \\[6pt]
\Delta\left(\frac{1}{n^2}\right)_4 \\[6pt]
\Delta\left(\frac{1}{n^2}\right)_5 \\[6pt]
\Delta\left(\frac{1}{n^2}\right)_6
\end{vmatrix}
=
\begin{vmatrix}
r_{11} & r_{12} & r_{13} \\[6pt]
r_{21} & r_{22} & r_{23} \\[6pt]
r_{31} & r_{32} & r_{33} \\[6pt]
r_{41} & r_{42} & r_{43} \\[6pt]
r_{51} & r_{52} & r_{53} \\[6pt]
r_{61} & r_{62} & r_{63}
\end{vmatrix}
\bullet
\begin{vmatrix}
F_x \\[6pt]
\\
F_y \\[6pt]
\\
F_z
\end{vmatrix}. \qquad (3.17)
$$

Matrix multiplication rules are used for Eq. (3.17). The 6×3 matrix with elements r_{ij} is known as the electro-optic tensor of the material. For LiNbO$_3$, the only non-vanishing elements r_{ij} are[3]

$$r_{33} = 30.8 \times 10^{-12}\,\text{m/V}, \quad r_{13} = 8.6 \times 10^{-12}\,\text{m/V}, \quad r_{22} = 3.4 \times 10^{-12}\,\text{m/V},$$

$$r_{42} = 28 \times 10^{-12}\,\text{m/V},$$

where

$$r_{13} = r_{23}, \quad r_{22} = -r_{12} = -r_{61}, \quad r_{42} = r_{51}.$$

Substituting these tensor elements into Eq. (3.17) and (3.16), we obtain a new index ellipsoid for a given $\boldsymbol{F}$

$$\left(\frac{1}{n_o^2} - r_{22}F_x + r_{13}F_z\right)x^2 + \left(\frac{1}{n_o^2} + r_{22}F_y + r_{13}F_z\right)y^2 + \left(\frac{1}{n_e^2} + r_{33}F_z\right)z^2$$
$$+ (2r_{42}F_x)zx + (-2r_{22}F_x)xy \quad = 1. \tag{3.18}$$

In principle, in order to determine how plane waves propagate under an applied $\boldsymbol{F}$ we need to find the new principal axes, x', y', and z' and then determine the polarization $\boldsymbol{D}_o'$, $\boldsymbol{D}_e'$ and n' values for the given $\boldsymbol{s}$.

For any incident radiation that has arbitrary polarization, its displacement field $\boldsymbol{D}$ *needs to be considered as the sum of ordinary and extra-ordinary wave polarizations,* $\boldsymbol{D}_o$ *and* $\boldsymbol{D}_e$, *at the input. Since these two plane waves have different phase velocities, their sum will have variable total polarization as they propagate. The difference between the propagation of plane waves with and without* $\boldsymbol{F}$ *represents the modulation produced by the electro-optic effect in an unbounded medium.*

3.1.2   Linear electro-optic effects in optical waveguides

Calculation of the electro-optic effect in guided-wave modes will be complex for a general $\boldsymbol{F}$ and for a waveguide oriented in an arbitrary direction. Fortunately, most commonly used optical waveguides are fabricated on a substrate along a specific crystal orientation. For example, LiNbO$_3$ waveguides are usually fabricated on x-cut substrates or z-cut substrates with direction of propagation in the y direction of the crystal. Figure 3.2 illustrates these two types of waveguide. Figure 3.2(a) shows a diffused waveguide on z-cut LiNbO$_3$. In order to take advantage of the large electro-optic coefficient, r_{33}, the $\boldsymbol{F}$ in the waveguide applied by the electrodes is predominantly in the z direction. For simplicity, let us assume for this discussion that $\boldsymbol{F}$ is uniform in the region occupied by the guided-wave mode. Therefore, $F_x = F_y \cong 0$, and the crystalline x, y, and z, axes are still the axes of the index ellipsoid with the applied $\boldsymbol{F}$. Plane waves propagating in the y direction with both $\boldsymbol{D}$ and $\boldsymbol{E}$ polarized in the z direction will have $n = n_e'$, while plane waves with x polarization of $\boldsymbol{D}$ and $\boldsymbol{E}$ will have $n = n_o'$. In these two polarizations, $\boldsymbol{D}$ is parallel and proportional to $\boldsymbol{E}$. According to Eq. (3.18), the n_e' and n_o' are

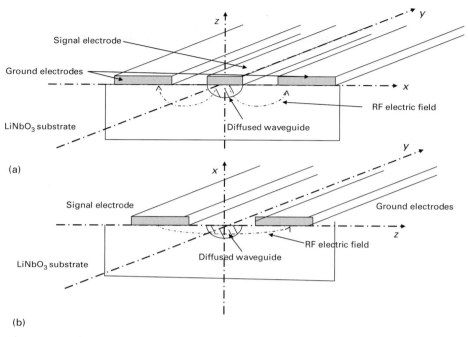

Fig. 3.2. Commonly used waveguide and electrode configurations in LiNbO$_3$. (a) A diffused waveguide on z-cut substrate. (b) A diffused waveguide on x-cut substrate. The direction of propagation is along the y direction. The RF field produced by the electrodes is oriented along the z direction in the core of the waveguide.

$$n_e' = \left(\frac{1}{n_e^2} + r_{33} F_z\right)^{-1/2} \cong n_e - \frac{1}{2} n_e^3 r_{33} F_z,$$

$$n_o' = \left(\frac{1}{n_o^2} + r_{13} F_z\right)^{-1/2} \cong n_o - \frac{1}{2} n_o^3 r_{13} F_z. \qquad (3.19)$$

For guided-wave modes propagating in the y direction, the TE modes have a dominant optical electric field polarized in the x direction for waveguides on z-cut substrates shown in Fig. 3.2(a), while the dominant optical electric fields in TM modes are polarized in the z direction. Note that $\underline{D}$ and $\underline{E}$ are parallel to each other for the dominant electric field in these two cases. Therefore, in the scalar approximation of the wave equation and for F uniform across the waveguide, the effective index of the TE modes can be calculated approximately by using n_o' in Eq. (1.4) and (1.5), while the effective index of TM modes can be calculated approximately by using n_e' in Eq. (1.26) and (1.27). A diffused waveguide on x-cut substrate is shown in Fig. 3.2(b). For waveguides along the y direction on x-cut substrates, the TE modes have the dominant electric field polarized in the z direction, and the TM modes have the dominant electric field polarized in the x direction. The F applied from the electrodes is predominantly in the z direction. In this case, the effective index of TE modes for uniform F can be calculated using n'_e, and the effective index of TM modes can be calculated using n'_o. In summary, in order to maximize the electro-optic effect, $\boldsymbol{F}$ is applied in the z direction to TE modes in x-cut LiNbO$_3$ and to TM modes in z-cut LiNbO$_3$. It is also clear that any change from these two cases, for example, an addition of F_x in addition to F_z may require us to find the x', y', and z' axes and then find the new $\boldsymbol{D}_e$ and $\boldsymbol{D}_o$ using Eq. (3.18). The analysis of the effective index of the guided modes would be much more complex.

For polymer waveguides discussed in Section 1.3.3 shown in Fig. 1.9, the direction in which the poling field is applied is usually defined as the z direction. The x and y axis are then in the plane parallel to the substrate. Material properties are symmetric in the x and y directions. The non-vanishing elements of the electro-optic tensor are $r_{13} = r_{23}$, $r_{42} = r_{51}$, and r_{33} [12]. The largest electro-optic coefficient is r_{33}. For example, it has been shown theoretically that $r_{33} = 3r_{13}$ [13]. Therefore, in order to maximize the electro-optic effect, $\boldsymbol{F}$ is usually applied in the z direction from the electrodes illustrated in Fig. 1.9. For such a configuration, the analysis of the electro-optic effect of TM modes is identical to that of the z-cut LiNbO$_3$ with $\boldsymbol{F} = Fi_z$ and a different r_{33} coefficient. On the other hand, the TE modes will not have an electro-optic effect. The value of the r_{33} coefficient will depend on polymer material engineering. The reported r_{33} is much larger than that of LiNbO$_3$, making the polymers very attractive for electro-optic applications. For example, $r_{33} = 130$ pm/V may be anticipated. In comparison, $r_{33} = 30.8$ pm/V in LiNbO$_3$. The challenge for polymer waveguide design is to obtain a material that has high glass temperature, low propagation loss and large electro-optic coefficient simultaneously [14].

Note that GaAs, InP, or other materials grown epitaxially on them have $r_{41} = r_{52} = r_{63}$, all other r_{ij}s are zero. In such a material with cubic crystalline symmetry, $n_x = n_y = n_z = n_o$. Therefore the equation of the index ellipsoid for all III-V compound semiconductor materials is [11]

$$\frac{x^2 + y^2 + z^2}{n_o^2} + 2r_{41}\left(F_x yz + F_y zx + F_z xy\right) = 1. \tag{3.20}$$

In GaAs, $n_o = 3.6$ and $r_{41} = 1.1 \times 10^{-12}$ m/V at the 0.9 μm wavelength, and $n_o = 3.3$ and $r = 1.43 \times 10^{-12}$ m/V at the 1.15 μm wavelength. Similar values of n_o and r_{41} have been reported in other III-V compound semiconductors (see Table 9.2 in [1]). As an example, for a RF electric field F in the z direction, we obtain

$$\frac{x^2 + y^2 + z^2}{n_o^2} + 2r_{41}F_z xy = 1. \tag{3.21}$$

Let $z'' = z$, $\sqrt{2}x'' = x + y$ and $\sqrt{2}y'' = -x + y$, then the index ellipsoid in x'', y'' and z'' is

$$\left(\frac{1}{n_o^2} + r_{41}F_z\right)(x'')^2 + \left(\frac{1}{n_o^2} - r_{41}F_z\right)(y'')^2 + \frac{1}{n_o^2}(z'')^2 = 1. \tag{3.22}$$

For a plane wave propagating along the y'' axis, the major and minor axes of the ellipse for $\mathbf{D}$ are the x'' and the z'' axes. Their n values are:
for the $\underline{\mathbf{D}}//\underline{\mathbf{E}}//z''$ axis

$$n = n_{\mathrm{o}},$$

for the $\underline{\mathbf{D}}//\underline{\mathbf{E}}//x''$ axis

$$n = n_{\mathrm{o}} - \frac{1}{2}n_{\mathrm{o}}^{3}r_{41}F_{z}.$$

For waveguides fabricated on z-cut substrates[4] and oriented in the y'' direction as shown in Fig. 3.3, the electric field is obtained by applying an electrical voltage across the i layer in the reverse biased p–i–n junction. Since the intrinsic layer is usually very thin, the electric field can be very high for a given voltage applied to the electrode. Let us assume again that the electric field is uniform in the intrinsic electro-optic layer. The effective indices of TE modes are found from Eq. (1.14) and (1.15), using $n_{\mathrm{o}} - \frac{1}{2}n_{\mathrm{o}}^{3}r_{41}F_{z}$ as the material index of the intrinsic layer. Note that TM modes have no electro-optic effect. Since Eq. (3.20) is symmetric in x, y, and z, this result is applicable to x-cut or y-cut samples with electric field applied in the x or y directions.

No matter what materials are used, nor the $\underline{\mathbf{F}}$ the waveguide structure, the electro-optic effect will create a Δn_{eff} of the guided-wave mode due to the $\Delta \underline{\mathbf{F}}$. After propagating a distance z, the Δn_{eff} produces a phase shift $\Delta \phi$ of the guided-wave mode where $\Delta \phi = \Delta n_{\mathrm{eff}} \omega \sqrt{\mu_{\mathrm{o}}\varepsilon_{\mathrm{o}}} z$.

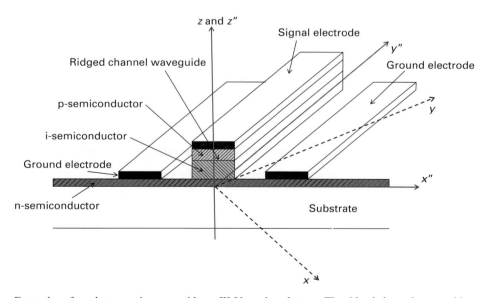

Fig. 3.3. Examples of an electro-optic waveguide on III-V semiconductors. The ridged channel waveguide on the z-cut substrate is oriented along the y'' direction which is 45° from both the x and the y axis. The high index intrinsic core of the waveguide is sandwiched between a p-doped and an n-doped semiconductor. A reverse bias is applied from the electrodes to the i layer through the p–i–n junction.

3.2   Electro-absorption effects in semiconductors

In semiconductors, electrons and holes are the particles that undertake stimulated emission and absorption. How such carriers are generated, transported and recombined has been discussed extensively in the literature [15–18]. We note in particular that they are in a periodic crystalline material. The energy levels of free electrons and holes are distributed in bulk semiconductors within conduction and valence bands. Within the conduction band and the valence band, each energy state has a wave function of the form [18]

$$\Psi_C(\underline{r}) = u_{C\underline{k}}(\underline{r})e^{j\underline{k}\bullet\underline{r}}, \tag{3.23}$$

where $u_{C\underline{k}}(\underline{r})$ has the periodicity of the crystalline lattice. The energy of electrons in the conduction band for a state with given $\underline{k}$ (known as the parabolic approximation of the energy band structure) is

$$E_e(|\underline{k}|) - E_C = \frac{\hbar^2|\underline{k}|^2}{2m_e}. \tag{3.24}$$

A similar expression is obtained for energy levels in the valence band,

$$E_h(|\underline{k}|) - E_V = -\frac{\hbar^2|\underline{k}|^2}{2m_h}. \tag{3.25}$$

Note that E_C is the bottom of the conduction band and E_V is the top of the valence band, m_e and m_h are respectively the effective mass of the electron and the hole. There are no energy levels between the conduction and valence band in pure bulk semiconductors. Note that $E_C - E_V$ is known as the bandgap E_{gap} of the material, $E_{gap} = E_C - E_V$. There are a large number of energy levels per unit energy range within each band, defined as the density of states. The specific distribution of energy levels (i.e. the m_e, the m_h and the parabolic approximation) depends on the material.

The probability that a given energy level is occupied by electrons (i.e. whether the level is empty or filled) depends on the carrier densities in the conduction band and in the valence band of the specific sample. When a photon with energy $h\nu$ is incident on the material, it may excite an electron from the valence band into the conduction band by stimulated absorption, i.e. creating simultaneously a hole in the valence band and an electron in the conduction band. Stimulated absorption takes place directly for a specific pair of energy levels only: (a) when the photon energy is equal to the difference of the energy levels in the conduction and in the valence band, i.e. $h\nu = E_e - E_h > E_C - E_V$; (b) when the energy level in the conduction band is empty; and (c) when the energy level in the valence band is filled with an electron (i.e. this energy level is not filled by a hole). The total probability of absorption at a given photon energy (i.e. wavelength) is the sum of the absorptions for all the energy levels with $E_e - E_h = h\nu$. It is proportional to the probability for stimulated transition between states, the density of states and the probability distribution of how the energy levels are occupied in the conduction band and the valence band. Since there are no energy levels between E_C and E_V, there is no direct transition for absorption by photons that have energy less than $E_C - E_V$.

3.2.1 The Frantz–Keldysh electro-absorption effect in bulk semiconductors

The Frantz–Keldysh effect denotes the electric field dependent optical absorption in bulk semiconductors [19, 20]. In basic semiconductor theory, when $hv < E_C - E_V$, the photon energy is not large enough directly to excite an electron from the valence band to the conduction band. However, experimental results and more detailed analyses show that the absorption spectra at the band edge exhibit a complicated structure, often giving an exponential tail in the long wavelength region, known as the Urbach's rule. There are many theories. The most convincing, attributed to Dow and Redfield, is that the exponential absorption tail is caused by electric field induced ionization of the exciton [21, 22]. This theory can be summarized as follows. Excitons are electron–hole pairs that are bound by the Coulomb potential with respect to each other. Thus the photon energy corresponding to the exciton transition will be slightly less than $E_C - E_V$. The ionization of the exciton into free electrons and holes is caused by an electric microfield of optical phonons, impurities and/or other mechanisms. Exciton transition and ionization will be affected by the applied electric field, thereby creating the electro-absorption effect. Although there is not a single theory that dominates, all the analyses showed a dependence of the exponential tail on the applied electric field. Figure 3.4 shows the absorption coefficient calculated according to Frantz for GaAs plotted as a function of the photon wavelength for several values of the electric field [23].

Clearly, in Fig. 3.4, at a wavelength slightly below the wavelength corresponding to the bandgap, the absorption coefficient is a function of the electric field. Therefore, the output intensity of an optical wave propagating through the material will be reduced, i.e. the χ'' will be increased, as the applied electric field is increased. This is the physical basis of Frantz–Keldysh electro-absorption modulation of optical radiation intensity. In principle, all III-V semiconductors have cubic symmetry, thus the Frantz–Keldysh effect should be independent of the direction of the applied electric field. On the other hand, since the probability of induced transition is proportional to the matrix element of electric dipole connecting the upper and the lower energy states, the absorption coefficient α due to the induced transition would depend on the polarization of the optical field.

Figure 3.5 shows the measured normalized transmitted radiation T as a function of electrical voltage V in an InGaAsP channel waveguide, 135 µm long, at 1.318 µm wavelength, and for the TE and TM modes of the waveguide [24]. In this case, the radiation photon energy is detuned from the band edge by 65 meV. Note that T as a function of V is clearly different for TE and TM polarizations. The normalized T is the ratio of the actual transmission to the transmission at $V = 0$. The relation between T and the change in absorption coefficient, i.e. $\Delta\alpha$ due to change in V, is shown in the figure where L is the length of the waveguide and Γ is the optical filling factor. The quantity Γ is the optical energy carried in the electro-absorption material as a fraction of the total optical energy carried by the waveguide mode. It is the same ($\Gamma = 0.66$) for TE and for TM polarization of the optical field. Therefore, data on $\Delta\alpha$ can be calculated directly from T.

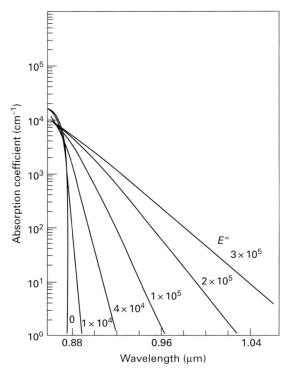

Fig. 3.4.   Calculated absorption coefficient of GaAs as a function of wavelength for several values of
electric field. Taken from ref. 23 with permission from *Applied Optics*.

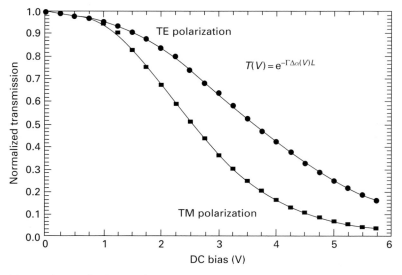

Fig. 3.5.   Measured normalized transmission of an InGaAsP channel waveguide in the TE and TM
polarizations. Taken from ref. 24 with permission from R. Welstand. The normalized transmission
is the ratio of the actual transmitted optical power to the optical power transmitted at zero applied
bias voltage.

Note that V is the electrical voltage applied to the reverse biased p–i–n structure in which the InGaAsP core constitutes the i layer. In other words, the applied electric field is proportional to V. Welstand has calculated that the applied field ΔF in the i layer corresponding to 4 V bias is 120 kV/cm. The $\Delta \alpha$ that has been achieved for the TM mode in this sample is approximately 200 cm^{-1} for a V variation from 3 to 5 V. Therefore, $\Delta F \cong 60$ kV/cm, and $\frac{\Delta \alpha}{\Delta F} \cong 3.3 \times 10^{-3}$ V^{-1}.

In order to use Frantz–Keldysh electro-absorption for an optical radiation at a given wavelength, it is obvious that one should use a semiconductor material with a bandgap energy, $E_C - E_V$, detuned slightly towards a slightly larger value from the photon energy of the radiation. From Fig. 3.4, it is clear that a radiation wavelength detuned approximately from the bandgap by 0.1 µm (i.e. $\Delta \lambda_{det} = 0.1$ µm) would be appropriate for GaAs.[5] In the InGaAsP waveguide sample shown in Fig. 3.5, $\Delta \nu_{det}$ is 64 meV. Since III-V semiconductors could be grown epitaxially on and lattice matched to a GaAs or InP substrate with a variety of bandgaps as shown in Fig. 1.8, it is straightforward to find a material that will provide Frantz–Keldysh electro-absorption at wavelengths important for optical fiber communication.

In all cases, electro-absorption creates a change of the transmittance, i.e. ΔT, of the intensity of the guided wave after a propagation distance L, by changing the absorption coefficient $\Delta \alpha$, i.e. $\Delta \chi''$, of the electro-absorption layer. However, the electro-absorption layer covers only a fraction of the lateral size of the guided-wave mode. Thus $\Delta T = \exp(-\alpha_o \Gamma L)[\exp(-\Delta \alpha \Gamma L)]$ where α_o is the absorption coefficient before the application of F, and Γ is the optical filling factor of the electro-absorption layer with respect to this mode.

3.2.2   Electro-absorption in quantum wells (QW)

3.2.2.1   Energy levels in quantum wells

A quantum well double heterostructure in semiconductors consists of a thin layer of material, called the well, that has a smaller bandgap, E_Γ, sandwiched between materials with a larger bandgap, E_g, called the barrier. These layers are typically III-V group semiconductors with different compositions that are grown epitaxially on and lattice matched to the GaAs or InP substrates as discussed in Section 1.3.2. The thickness of the well L_W is typically 50 to 150 Å. The barrier is just thick enough (e.g. 50 to 100 Å) to isolate the wells. Figure 3.6(a) shows a typical one-dimensional potential diagram of the conduction and valence bands as a function of thickness position x at zero applied electric field. At the interface of the well and the barrier, there are discontinuities in conduction band edge ΔE_C and valence band edge ΔE_V, $\Delta E_C + \Delta E_V = E_g - E_\Gamma$. Quantum mechanical calculations of energy states in such potential wells yield discrete energy levels E_e for electrons in the conduction band and discrete energy levels E_h for holes in the valence band [25]. In the example illustrated in Fig. 3.6, E_{e1} is the lowest order energy level for electrons in the conduction band. The energy state for this energy level is illustrated as ψ_e. Some holes in the valence band have a greater mass, called heavy holes, and some holes have a lesser hole mass, called light holes. The highest hole energy in the valence band is usually for heavy holes. Therefore only E_{hh1} and its

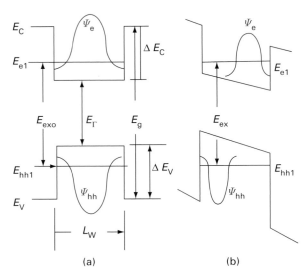

Fig. 3.6.   Potential energy diagrams, energy levels and energy states in quantum wells. (a) At zero electric
field. (b) At a bias electric field. (The figure is taken from Fig. 6.6 of *RF Photonic Technology
in Optical Fiber Links*, ed. W. S. C. Chang, Cambridge University Press, 2002.)

energy state ψ_{hh} are illustrated in Fig. 3.6(a). Other higher electron levels and lower
hole levels such as E_{e2} and E_{lh1} are not shown here. The energy states ψ demonstrate
that electrons and holes in a quantum well are confined in the x direction. A multiple
quantum well (MQW) structure consists simply of multiple periods of quantum wells
separated by barriers.

3.2.2.2 Exciton transitions and absorption

Energy levels of the electrons and holes in the thickness, x, direction are E_e and E_h. The
total energy of electrons and holes is the sum of their energy in the x direction, i.e. $E_e +
E_h$, and the energy of an electron–hole pair in the yz plane, E_{yz}. In order to understand
the energy of the electron–hole pair in the yz plane, let us consider first the energy of
electron–hole pairs in three dimensions in bulk semiconductors. When electron–hole
pairs are created by absorption of a photon, they are initially closely spaced. In bulk
semiconductors, such electron and hole pairs will experience mutual three-dimensional
Coulomb forces similar to those present in a hydrogen atom. The energy of such an
electron–hole pair is lower than the energy of free electrons and holes; this electron–
hole coordination gives rise to a set of energy levels (called exciton levels) just below
the bandgap. The exciton spectra in bulk materials have been directly observed only at
very low temperatures. These excitons are responsible for the Urbach tail and the
Frantz–Keldysh effect discussed in the preceding section. The situation is different in
quantum wells. In the y and z directions of the quantum well, electrons and holes are
also subject to periodic potentials and forces in a bulk crystal. However, the quantum
confinement in the x direction increases the binding energy of the exciton. In the limit of
a two-dimensional exciton, the binding energy is four times the binding energy of

a three-dimensional exciton in bulk material. The binding energy of excitons in quantum wells is typically less than 15 meV.

Exciton absorption in quantum wells has been observed directly at room temperatures and under applied electric field. The stimulated transition of heavy hole excitons takes place at photon energy just below $E_{e1} - E_{hh1}$. The solid curve in Fig. 3.7 shows the TE polarized absorption spectrum of an $InAs_{0.4}P_{0.6}$(93 Å thickness wells)/$Ga_{0.13}In_{0.87}P$(135 Å thickness barriers) multiple quantum well (MQW) at zero applied electric field [26]. The heavy hole exciton transition has a transition wavelength shown as λ_{exo} with a linewidth δ_{exo}. For this sample, the half width half maximum δ_{exo} is 6 meV. A second transition with a less distinct absorption peak due to a light hole may also be seen in this figure at $\lambda = 1.250\,\mu m$. Since the photon energy corresponding to exciton transition is very close to $E_e - E_h$, the photon energy (or wavelength) for exciton transition can be estimated in practice by just calculating $E_e - E_h$. For example, $E_{e1} - E_{hh1}$ can be calculated from E_Γ, ΔE_C and ΔE_V to estimate the wavelength for a heavy hole exciton, and $E_{e1} - E_{lh1}$ can be calculated for a light hole exciton.

Note that the absorption coefficient α will be dependent on the polarization of the electric field because the matrix element for any induced transition between an electron and a hole is polarization dependent. For a plane optical wave propagating along the x direction, its electric field is polarized in the yz plane; its absorption coefficient α will be the same as a TE guided-wave mode propagating along any direction in the yz plane. For TM modes in a waveguide oriented in the y plane, its dominant electric field will be in the x direction, and its α will be different.

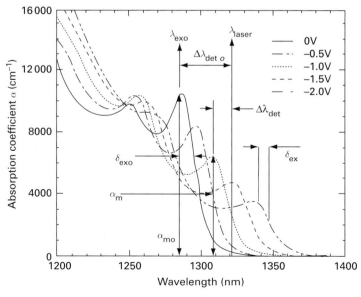

Fig. 3.7.   Absorption spectra of InAsP/GaInP multiple quantum well at different bias voltages. (The figure is taken from Fig. 6.3 of *RF Photonic Technology in Optical Fiber Links*, ed. W. S. C. Chang, Cambridge University Press, 2002.)

3.2.2.3   The quantum confined Stark effect (QCSE)

Under the application of an electric field in the x direction, the potential wells are tilted as shown in Fig. 3.6(b). The quantum mechanical solution for the energy values in the quantum well indicates usually a reduction in $E_{e1} - E_{hh1}$. Therefore, the exciton absorption line at E_{exo} shifts normally toward longer wavelengths (i.e. absorption peak at smaller photon energy), known as a red shift. Occasionally the shift in a specific potential well configuration may be toward a shorter wavelength, known as a blue shift. This is the quantum confined Stark effect (QCSE) [27–29]. Note also that as the potential wells are tilted the wave functions of energy states for electrons and for holes are also shifted to the opposite side of the quantum well as illustrated in Fig. 3.6(b). Since the amplitude of the stimulated absorption between the two energy states depends on the matrix element of the electric dipole connecting ψ_e and ψ_{hh}, the shift of energy state function to the opposite side of the quantum well will produce a reduction of the exciton absorption as the electric field is increased.

The QCSE, the reduction in the absorption coefficient at the exciton peak α_m and the broadening of the exciton line width δ_{ex} are clearly demonstrated in Fig. 3.7 as the applied voltage is increased. In this case, the electric voltage shown in the figure is applied across a reversed biased p–i–n junction that has an i layer approximately 0.5 μm thickness (containing 21 periods of quantum wells and barriers). Therefore the electric field F in units of V/cm applied to the quantum well is approximately 2×10^4 times the applied voltage. In this figure, the laser radiation at the wavelength λ_{laser} is detuned from the exciton peak at $F = 0$ by $\Delta\lambda_{deto}$. As the QCSE increases, the absorption coefficient in the MQW for the λ_{laser} shown in Fig. 3.7 will first increase when $\Delta\lambda_{det} > 0$ and then decrease when $\Delta\lambda_{det} < 0$ as F is increased. When the electric voltage is changed from 0.5 to 1.5 V, the change in absorption coefficient at λ_{laser} shown in the figure is $\Delta\alpha \cong 4000$ cm. Thus $\Delta\alpha/\Delta F \cong 200 \times 10^{-3}$ V^{-1}.

Figure 3.8 shows the measured QCSE and the calculated shift of $E_{e1} - E_{hh1}$ of the sample used in Fig. 3.7. The discrepancy has been attributed to the variation of the exciton binding energy as the applied electric field is varied. Figure 3.9 illustrates the $\Delta\alpha$ at the different detuning energies and reverse biases that can be obtained in this sample. Note the importance of small δ_{exo} and appropriate choice of detuning energy and reverse bias in order to maximize $\Delta\alpha$ for a given ΔF.

Quantum well structures became a reality because epitaxy technology provided the means for control of the layer thickness and smoothness to atomic level accuracy. Quantum wells and barriers are always parallel to the surface of the substrate. In other words, the thickness direction in the QW and the direction of the applied electric field must be perpendicular to the substrate surface. The most effective way to apply such an electric field is by fabricating a p–i–n structure parallel to the substrate surface where the MQW constitutes the i layer. In a reverse biased p–i–n structure, the electrical field is predominantly perpendicular to and focused in the i layer. This is the way in which α is obtained in Fig. 3.7, 3.8, and 3.9. It is the α for the TE polarization. Note that the measured α is the averaged absorption coefficient of the entire MQW layer. For a given

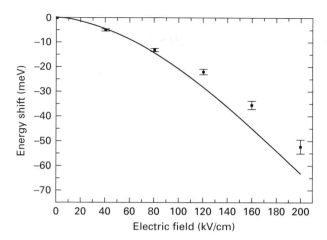

Fig. 3.8.   Quantum confined Stark effect of the $E_{e1} - E_{hh1}$ transition versus the electric field. Taken from ref. 25 with permission from X. Mei. The scattered signs are experimental data, and the solid curve is calculated theoretically using the effective width model.

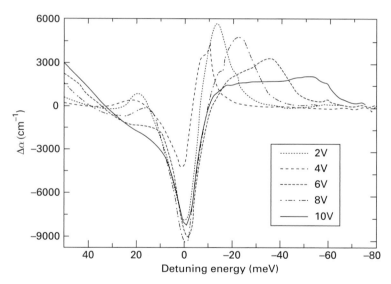

Fig. 3.9.   Change of absorption coefficient at different detuning energy and reverse biases. Taken from ref. 25 with permission from X. Mei.

electric field, the actual absorption takes place only in the well, not in the barrier. Therefore $\Delta\alpha/\Delta F$ is increased by using a thinner barrier layer. The minimum barrier thickness will be governed by the decay of the energy state functions, ψ_e and ψ_h, in the barrier. The conventional guideline is that the barriers be thick enough to ensure that the energy states of adjacent wells will not significantly interact with each other.

3.2.2.4   Characterizing the QW structures and the electro-absorption effect

Note that the sharper the exciton absorption line at $F = 0$ (i.e. the smaller the δ_{exo}), the smaller the required detuning $\Delta\lambda_{det}$, and the larger the $\Delta\alpha/\Delta F$. Note that δ_{exo} is determined in practice mostly by the quality of the QW layers, such as interface roughness between the well and the barrier, the defect density or the uniformity. Ultimately, when the material quality is excellent, δ_{exo} will be limited by the phonon broadening process. The crystalline quality of the material, grown epitaxially, can be examined by double axis high resolution X-ray diffraction techniques. High resolution transmission electron microscopy can be used to examine any local atomic displacements at and around interfaces. At low temperatures, phonon broadening will be negligible. The quality of the material can be evaluated by observing the linewidth of photo-luminescence spectra of the exciton transition at temperatures below 10 K. Finally, the absorption coefficient can be measured directly from the sample using plane waves propagating perpendicularly to the QW layer in TE polarization and guided waves propagating in waveguides parallel to the QW layers in TE or TM polarization.

3.2.2.5   Saturation of QW electro-absorption

Absorbed light generates electron–hole pairs in quantum wells. The separated electrons and holes are swept out of the quantum wells by the bias electric field. Holes have a slower escape rate because of the larger mass. Holes may be trapped in the quantum well heterostructure interfaces due to the valence band discontinuity ΔE_V. As the photo-generated holes accumulate, an opposing space charge electric field builds up and perturbs the applied electric field. Under high optical power illumination, the photo-generated space charge densities may be large enough to redistribute the electric field and to reduce the electro-absorption effect. In other words, QW electro-absorption is known to saturate at high optical power [30].

3.2.3   Comparison of Frantz–Keldysh and QW electro-absorption

The objective of all electro-absorption devices is to obtain the largest change in waveguide absorption of the optical power using the smallest change in F. Clearly the principal advantage of using quantum well structures for electro-absorption is that the $\Delta\alpha/\Delta F$ is much larger in quantum wells than in the bulk. However, the Frantz–Keldysh effect can be obtained from an electric field applied in any direction, while QCSE in quantum wells can be obtained only with electric field applied in the direction perpendicular to the quantum well layers. It means that there are more options in the design of electrodes with the Frantz–Keldysh effect. This may be significant when one considers the electrical response of the devices. In addition, in order to avoid saturation, photo-generated carriers created by absorption must be swept away by a bias electric field. When there is band edge discontinuity between the well and the barrier, it is easier to retain carriers, and thereby easier to saturate in quantum well structures. Since there is no valence band discontinuity in bulk material, the Frantz–Keldysh electro-absorption effect saturates at much higher optical power. Bulk materials can also be grown by the liquid phase epitaxy (LPE) technique allowing us more alternatives in fabricating the waveguide, using techniques such as etching and regrowth.

3.3 The electro-refraction effect

The χ' and the χ'' of any passive material are related mathematically to each other through the Kramers–Kronig relation

$$\chi'(\omega) = \frac{1}{\pi}\mathrm{P.V.}\int\limits_{-\infty}^{+\infty} \frac{\chi''(\omega')}{\omega' - \omega}\,\mathrm{d}\omega',$$

$$\chi''(\omega) = -\frac{1}{\pi}\mathrm{P.V.}\int\limits_{-\infty}^{+\infty} \frac{\chi'(\omega')}{\omega' - \omega}\,\mathrm{d}\omega', \tag{3.26}$$

where P.V. stands for the Cauchy principal value of the integral that follows [31]. It is a mathematical result based on a contour integration of $\int_c \frac{\chi(\omega')}{\omega'-\omega}\mathrm{d}\omega' = 0$ in the lower half ω' plane for any χ that does not have a pole in the lower half plane. In essence, whenever there is a change in absorption spectra there is a corresponding change of the spectra of the refractive index. Therefore, there are refractive index changes accompanying any electro-absorption.

For guided waves propagating in the z direction, the electric and magnetic fields have $\exp(\mathrm{j}\omega t - \mathrm{j}\beta z)$ variation. For guided waves with attenuation, the propagation wave number β is also complex, $\beta = \beta' - \mathrm{j}\beta''$, $\beta' = n_{\mathrm{eff}}\,\omega/c$. Note that β'' is related to the intensity absorption coefficient α by $\beta'' = \alpha/2k$. Since Eq. (3.26) is derived mathematically for any χ of a passive material that does not have a pole in the lower half plane of complex ω', the Kramers–Kronig relation applies equally well to β' and β'' of any passive guided-wave material structure. Thus

$$n_{\mathrm{eff}} = \frac{1}{\pi}\mathrm{P.V.}\int\limits_{-\infty}^{+\infty} \frac{(\alpha c/2\omega)}{\omega' - \omega}\,\mathrm{d}\omega',$$

$$\frac{\alpha c}{2\omega} = -\frac{1}{\pi}\mathrm{P.V.}\int\limits_{-\infty}^{+\infty} \frac{n_{\mathrm{eff}}}{\omega' - \omega}\,\mathrm{d}\omega'. \tag{3.27}$$

A relationship identical to Eq. (3.27) also exists between Δn_{eff} and $\Delta\alpha$ generated by electro-absorption. *In summary, in any material such as a multiple quantum well structure, one can choose whether to use the Δn_{eff} or $\Delta\alpha$ to achieve the modulation. In electro-absorption modulation, material structure and voltage bias are designed to yield large $\Delta\alpha/\Delta F$ at a specific optical wavelength. The accompanying Δn_{eff} causes a phase shift of the guided wave that is the cause of chirping in electro-absorption modulators (see Section 6.2.1.4). In electro-refraction modulators, material structure and voltage bias are designed to achieve large $\Delta n_{\mathrm{eff}}/\Delta F$ while the α is kept within certain limits at the specific wavelength. Usually, the $\Delta n_{\mathrm{eff}}/\Delta F$ volume that can be achieved in the electro-refraction effect is much larger than those in the electro-optic effect.* Comparing the electro-refraction to the electro-optic effect, it is clear that there will always be some

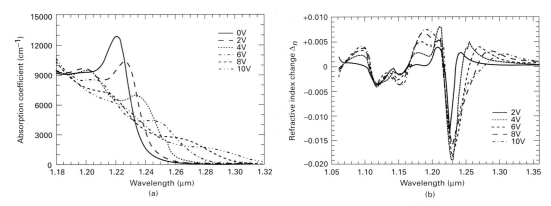

Fig. 3.10.   Electro-absorption and electro-refraction spectra of an InAsP/GaInP multiple quantum well. Taken from ref. 32 with permission from K. Loi. (a) Measured absorption spectra at various biases. (b) Calculated change of index as a function of wavelength for various changes of bias.

absorption in materials exhibiting the electro-refraction effect, while the electro-optic effect creates no absorption.

Figure 3.10(a) shows the measured TE absorption α spectra at room temperature at various reversed bias for a 30 period InAsP(well)/GaInP(barrier) MQW sample grown by gas source MBE on an InP substrate [32]. The exciton peak of the material is at the 1.22 μm wavelength. A QCSE of 46 meV has been observed at 10 V bias. Figure 3.10(b) shows the Δn calculated by Eq. (3.27) as a function of wavelength at various bias voltages. Since $\Delta\alpha$ was measured over only a finite spectral range, $\Delta\alpha$ was assumed to be zero outside the measurement wavelength range in the calculation for Δn. A comparison of Δn and $\Delta\alpha$ shows that Δn rolls off much slower than $\Delta\alpha$ at the longer wavelength region. For this reason, if electro-refraction is used for modulation it will be more effective to use a larger detuning energy (e.g. 60 meV), so that there will be a large $\Delta n/\Delta F$ while the α will be kept small at the operating wavelength. Small α is necessary so that the attenuation of the optical intensity will be reasonable over the length of the waveguide. In comparison, for electro-absorption modulation, the detuning energy needs to be much smaller (e.g. 30 meV) in order to obtain a large $\Delta\alpha/\Delta F$. For this sample, Δn of 2×10^{-3} was predicted for a 10 V change of bias voltage at the 1.32 μm wavelength while a small α is maintained.

3.4   The acousto-optical effect

Mechanical strain in a solid causes a change in the index of refraction. This photo-elastic effect is characterized by a photo-elastic tensor that relates the strain tensor to the optical refractive indices. In the case of the acousto-optic effect, mechanical strain (i.e. the index change) is produced by an acoustic wave. Usually optical waveguides are only a few μm thick. Thus surface acoustic waves are used to create the acoustic-optic effect corresponding to the thickness of a waveguide.

Let there be an acoustic surface wave propagating in the y direction. The net effect of the acoustic wave is to create a periodic traveling wave of $\Delta\varepsilon$ in that direction. Mathematically, a simplified $\Delta\varepsilon$ caused by an acoustic wave is described as

$$\Delta\varepsilon(x,y,z,t) = \Delta\varepsilon \cos\left(\mathbf{\underline{K}_a} \cdot \boldsymbol{\rho} - \Omega t\right) \text{rect}\left(\frac{t/2+x}{t/2}\right) \text{rect}\left(\frac{W/2-z}{W/2}\right),$$

$$\mathbf{\underline{K}_a} = K_a \underline{i}_y, \quad \boldsymbol{\rho} = y\underline{i}_y + z\underline{i}_z, \quad \varepsilon = n^2 \varepsilon_o, \quad \Delta\varepsilon = (2n\Delta n)\varepsilon_o. \tag{3.28}$$

Here, ε_o is the free space electric permittivity; Ω is the frequency of the acoustic wave; $\mathbf{\underline{K}_a}$ is the vector representation of the propagation wave number for the acoustic wave; W is the width of the acoustic wave in the z direction; t is the height of the acoustic wave in the x direction; $\boldsymbol{\rho}$ is the coordinate vector; and Ω/K_a is the acoustic velocity v_{ac}. The rectangular functions, $\text{rect}(u)$,[6] designate an acoustic wave confined to the layer from $x = 0$ to $x = -t$ and within a width W.

The change in the refractive index Δn is related to the strain by

$$\Delta\left(\frac{1}{n^2}\right) = pS, \tag{3.29}$$

where p is the photo-elastic constant; and S is the strain amplitude of the acoustic wave. Note that Δn is related to the power of the acoustic wave, P_{ac} [33], by

$$\Delta n = \sqrt{\frac{n^6 p^2 10^7 P_{ac}}{2\rho_m v_{ac}^3 t W}} = \sqrt{M_2 10^7 P_{ac}/2tW}. \tag{3.30}$$

Here P_{ac} is the total acoustic power in watts; v_{ac} is the acoustic velocity; and ρ_m is the mass density of the material. In crystalline solids, p depends strongly on the orientation. Usually, Δn that can be obtained from surface acoustic waves is small even for optimum choice of material and orientation. For example, $M_2 = 6.9 \times 10^{-18}\,\text{s}^{-3}/\text{g}$ in LiNbO$_3$. A phase matched interaction between an optical guided wave and an acoustic wave, like those discussed in Section 2.2.3 for interaction between a grating and a guided wave, needs to be used to produce significant deflection or switching effect (see Sections 3.5.3 and 5.5, and [34]).

Surface acoustic waves are usually generated by fabricating a set of interdigital electrodes on piezoelectric material, or by bonding an acoustic transducer to the material [35]. Figure 3.11 is a schematic of a simple interdigital acoustic surface wave transducer on a piezoelectric substrate. When a RF voltage is applied to the electric port, a fringing electric field is excited between the fingers of the electrodes. Consequently, the polarity of the electric field alternates from one electrode to the next, exciting the acoustic wave via the piezoelectric effect. The width of the acoustic wave generated is W. The electrode spacing is L. At the center acoustic frequency f_o, $L = \Lambda_{ac}/2 = v_{ac}/2f_o$ where Λ_{ac} is the acoustic wavelength at f_o. Acoustic power output versus frequency has the form

$$P_{ac}(f) = \left[\frac{\sin(N\pi(f-f_o)/f_o)}{\sin(\pi(f-f_o)/f_o)}\right]^2. \tag{3.31}$$

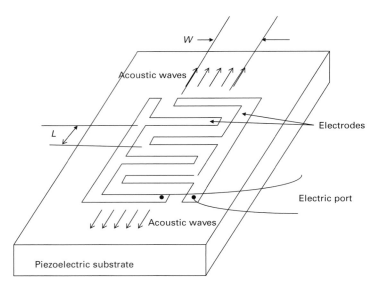

Fig. 3.11.   An interdigital electrode, surface acoustic wave transducer. (See ref. 35 for more details.)

Clearly there is only a finite bandwidth within which an acoustic wave can be efficiently generated. There is also a conversion efficiency in which the RF electrical drive power is converted into P_{ac}. Thus, there is a bandwidth and an efficiency for generation of the acoustic wave. More sophisticated design could provide trade-offs between efficiency and bandwidth. In addition, the attenuation of acoustic waves would increase at higher f_o, thereby limiting the effective frequency range in which the acousto-optic effect can be used for deflection and switching of optical guided waves [36].

3.5   A perturbation analysis of electro-optical effects

No matter how an electro-optical effect is created (by electro-optic, electro-absorption, electro-refraction and acousto-optic effects), the propagation of the guided-wave mode (i.e. the n_{eff} or attenuation) is affected by a change in susceptibility of the material. In the beginning of this chapter we showed how the propagation of a plane wave is affected by a uniform change of susceptibility. However, the electro-optic change of index is not uniform over the entire guided wave. Perturbation technique analysis presented in Section 2.1 could be used to evaluate conveniently the effect of the change in susceptibility in the electro-optical active region on the propagation of the guided-wave mode. Three examples will be presented in this section, perturbation of the effective index by $\Delta\chi'$, attenuation of guided wave by $\Delta\chi''$, and scattering of a planar guided wave by an acoustic wave.

Let the nth unperturbed guided-wave mode be $\underline{e_n}$ with amplitude a_n. The electro-optical effect creates a perturbation $\Delta\varepsilon$ only in the active region of the material structure

$$\Delta\varepsilon = (\Delta\chi' - j\Delta\chi'')\varepsilon_0. \tag{3.32}$$

Following the notation used in Eq. (3.1) to (3.4), and for an active material with an unperturbed refractive index n, the electro-optical effect creates a new ε

$$\varepsilon = (\chi_0 + \Delta\chi)\varepsilon_0 = (n^2 + \Delta\chi')\varepsilon_0 - j\Delta\chi''\varepsilon_0. \tag{3.33}$$

3.5.1 Perturbation of the effective index n_{eff} by $\Delta\chi'$

When the refractive index in the active region is changed from n to $n + \Delta n$,

$$\Delta\chi' = 2n\Delta n. \tag{3.34}$$

From Eq. (2.6),

$$\frac{da_n}{dz} = -ja_n C_{n,n}, \tag{3.35}$$

$$C_{n,n} = \frac{\omega}{4} \iint_{\text{active region}} \Delta\varepsilon \left(\underline{e_n} \bullet \underline{e_n^*} \right) ds. \tag{3.36}$$

Therefore, when $\Delta\varepsilon$ is independent of z

$$a_n = A e^{-j\Delta\beta z}, \qquad \Delta\beta = \beta_0 \Delta n_{\text{eff}} = \frac{\omega n \varepsilon_0}{2} \int_{\text{active region}} \Delta n \left(\underline{e_n} \bullet \underline{e_n^*} \right) ds. \tag{3.37}$$

When the active region covers the entire guided-wave mode, Eq. (3.37) reduces to a result like the one obtained in Eq. (3.8) for plane waves that experience a uniform electro-optic effect. Since the $\underline{e_n}$ is normalized according to the modal analysis in Chapter 1 and Eq. (2.3),

$$\frac{\beta}{2\omega\mu} \iint_{\infty} \underline{e_n} \bullet \underline{e_n^*} ds = 1,$$

we obtain

$$\Delta\beta = \Delta n_{\text{eff}} \beta_0 = \beta_0 \frac{n}{n_{\text{eff}}} \Gamma \Delta n_{\text{av}}, \tag{3.38}$$

where

$$\Gamma = \frac{\displaystyle\iint_{\text{active region}} \Delta n \left(\underline{e_n} \bullet \underline{e_n^*} \right) ds}{\Delta n_{\text{av}} \displaystyle\iint_{\infty} \underline{e_n} \bullet \underline{e_n^*} ds}. \tag{3.39}$$

Note that Δn_{av} is the average Δn taking place in the active region, Γ is a filling factor for the Δn created by the electro-optical effect. The electro-optical phase shift for a guided wave propagating over a distance L is $\Delta\phi = \Delta n_{\text{eff}} \beta_0 L$. Similar expressions can be obtained for the TM modes.

3.5.2   Attenuation of guided-wave mode by $\Delta\chi''$

When there is $\Delta\chi''$ caused by electro-absorption, the effect of such perturbation on the guided-wave mode could also be calculated by perturbation analysis. According to Eq. (2.6), we have

$$\frac{da_n}{dz} = -\frac{\Delta\alpha_n}{2}a_n, \qquad \frac{\Delta\alpha_n}{2} = \frac{\omega}{4}\varepsilon_0 \underset{\text{active region}}{\iint} \Delta\chi''\left(\underline{e_n}\bullet\underline{e_n^*}\right)ds. \tag{3.40}$$

The solution of Eq. (3.40) is

$$a_n = Ae^{-\frac{\Delta\alpha_n}{2}z}. \tag{3.41}$$

Similarly, in view of the normalization of $\underline{e_n}$ we can rearrange the expression for $\Delta\alpha_n/2$ of TE modes into the form

$$\frac{\Delta\alpha_n}{2} = \frac{\Gamma\Delta\alpha_{av}}{2} = \frac{\beta_0}{2n_{eff}}\Gamma\Delta\chi''_{av}, \qquad \Gamma = \frac{\underset{\text{active region}}{\iint}\Delta\chi''\left(\underline{e_n}\bullet\underline{e_n^*}\right)ds}{\Delta\chi''_{av}\underset{\infty}{\iint}\underline{e_n}\bullet\underline{e_n^*}ds}, \tag{3.42}$$

where $\Delta\chi''_{av}$ is the average $\Delta\chi''$ in the active region and Γ is a filling factor of $\Delta\chi''$ in the active region. This result reduces again to that in Eq. (3.10) when the active region covers more than the guided-wave mode. Note that according to Eq. (3.42), the transmitted intensity of the guided wave after propagating a distance L is $T = \exp(-\alpha_0 L)\exp(-\Gamma\Delta\alpha_{av}L)$. Note that α_0 includes all the residual attenuation that existed in the absence of the modulation electric field. Similar expressions can again be obtained for TM modes.

3.5.3   The diffraction of a planar guided wave by acoustic surface waves

Let there be TE_0 planar guided waves in different directions in the yz plane. Each guided wave is designated by the angle θ_j that its direction of propagation makes with respect to the z axis. For small θ_j, the electric field of the TE_0 modes is still polarized in the y direction. Therefore the total field of a summation of TE modes can be expressed mathematically as

$$E_y'\underline{i_y} \cong \left[\sum_j a_j e^{-jn_0 k_j \cos\theta_j z}e^{-jn_0 k_j \sin\theta_j y}\right]E_{0,y}(x)e^{j\omega_j t}\underline{i_y}$$

$$\cong \left[\sum_j a_j e^{-j\underline{\beta_j}\bullet\underline{\rho}}\right]E_{0,y}(x)e^{j\omega_j t}\underline{i_y}, \tag{3.43}$$

where

$$\underline{\beta_j} = n_{eff}k_j\left(\cos\theta_j\underline{i_z} + \sin\theta_j\underline{i_y}\right), \quad \text{and} \quad \underline{\rho} = z\underline{i_z} + y\underline{i_y}.$$

Note that $E_{0,y}(x)$ describes the x variation of the TE$_0$ mode. Note that, in anticipation of the traveling acoustic wave interaction which will couple incident and diffracted waves at slightly different frequency, we have allowed the guided-wave modes to be at slightly different frequencies.

Let us consider two specific planar TE$_0$ guided-wave modes, propagating in the directions $+\theta$ (for $\underline{\beta}_d$ of the deflected wave) and $-\theta$ (for $\underline{\beta}_i$ of the incident wave) with respect to the z axis. The complex amplitudes for these modes are a_d and a_i. In this case, the acoustic $\Delta\varepsilon$ couples the incident wave, a_i, to the diffracted wave, a_d, as shown in Fig. 3.12. Equation (2.6), modified by the frequency variations of the incident and deflected wave, is directly applicable to a_i and a_d. For the incident and the deflected modes, we obtain

$$\frac{da_i}{dz} = -ja_d C_a e^{+j\left(\underline{\beta}_i - \underline{\beta}_d\right)\cdot\underline{\rho}} e^{j(\omega_d - \omega_i)t}\left[e^{j\underline{K_a}\cdot\underline{\rho}}e^{-j\Omega t} + e^{-j\underline{K_a}\cdot\underline{\rho}}e^{j\Omega t}\right],$$

$$\frac{da_d}{dz} = -ja_i C_a e^{-j\left(\underline{\beta}_i - \underline{\beta}_d\right)\cdot\underline{\rho}} e^{j(\omega_i - \omega_d)t}\left[e^{j\underline{K_a}\cdot\underline{\rho}}e^{-j\Omega t} + e^{-j\underline{K_a}\cdot\underline{\rho}}e^{j\Omega t}\right],$$

$$C_a = \frac{\omega}{4}\Delta\varepsilon_0 \int\limits_{-t}^{0} \left|\underline{e}_0 \cdot \underline{e}_0^*\right| dx. \tag{3.44}$$

Clearly the phase matching condition for maximum interaction between a_i and a_d is

$$\left(\underline{\beta}_i - \underline{\beta}_d\right) \cdot \underline{\rho} = \mp\underline{K_a} \cdot \underline{\rho} \quad \text{or} \quad \underline{\beta}_d = \underline{\beta}_i \pm \underline{K_a}. \tag{3.45}$$

This is known as the Bragg condition for acousto-optical deflection. The phase matching condition expressed in Eq. (3.45) is a vector relation in the yz plane. The phase matching

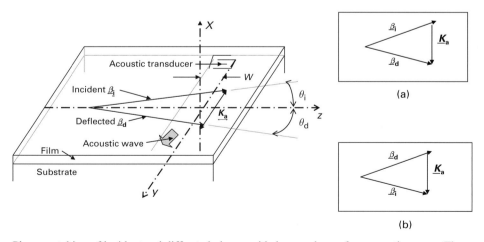

Fig. 3.12.   Phase matching of incident and diffracted planar guided waves by surface acoustic waves. The matching of the propagation wave vectors, $\underline{\beta}_i$ and $\underline{\beta}_d$, of the incident and diffracted planar guided waves in the θ_i and θ_d directions by the surface acoustic waves with $\underline{K_a}$ propagation vector. The width of the acoustic wave is W, which is also the distance of acousto-optic interaction. (a) $\omega_d = \omega_i + \Omega$. (b) $\omega_d = \omega_i - \Omega$.

condition in the z direction is satisfied independent of the K_a value because of the balanced $+\theta$ and $-\theta$ orientations of the $\boldsymbol{\beta}$s, and $|\boldsymbol{\beta_i}| = |\boldsymbol{\beta_d}| = n_{eff}k$. Here, in anticipation that $\omega_i \cong \omega_d$, we have taken the approximation $k_i = k_d = k$. Clearly, the magnitude of K_a determines the angular relationship between $\boldsymbol{\beta_i}$ and $\boldsymbol{\beta_d}$, i.e. the θ. In addition, according to Eq. (3.44), the interaction is strong only when

$$\omega_d = \omega_i \mp \Omega. \tag{3.46}$$

Since Ω (in RF frequency) $<< \omega_i$ and ω_d (in optical frequencies), $k_i = k_d = k$. The case using the upper sign in Eq. (3.45) is illustrated in inset (a) of Fig. 3.12. The case using the lower signs in Eq. (3.45) is illustrated in inset (b). Notice that the diffracted wave is at a slightly different optical frequency than the incident wave. Such a method is sometimes used to shift the optical frequency slightly from ω_i to ω_d.

When the phase and frequency matching conditions are satisfied, the solution to Eq. (3.44) is now a $\cos(C_a z)$ or $\sin(C_a z)$ variation. The exact form of the solution will again depend on the boundary conditions. Let a_i be the amplitude of the incident wave and a_d be the amplitude of the diffracted wave. The interaction by the acoustic wave begins at $z = 0$ and ends at $z = W$. Thus, the boundary condition is "$a_i = A$ and $a_d = 0$ at $z = 0$". For this boundary condition and for the case shown in Fig. 3.12, the solution of the amplitude of the two planar guided waves is

$$a_i(z) = A\cos(C_a z), \qquad a_d(z) = -jA\sin(C_a z). \tag{3.47}$$

The power diffraction efficiency, $|a_d(z = W)/a_i(z = 0)|^2$, and the power transmission efficiency, $|a_i(z = W)/a_i(z = 0)|^2$, are

$$\left|\frac{a_d(W)}{a_i(0)}\right|^2 = \sin^2(C_a W), \qquad \left|\frac{a_i(W)}{a_i(0)}\right|^2 = \cos^2(C_a W),$$

$$\left|\frac{a_d(W)}{a_i(0)}\right|^2 + \left|\frac{a_i(W)}{a_i(0)}\right|^2 \equiv 1. \tag{3.48}$$

For applications such as the acousto-optical switch or optical frequency shifter, maximum diffraction efficiency is desired. In that case, we need $W = \pi/2C_a$. For devices which require only low efficiency acousto-optical diffraction, the fraction of the optical power diffracted into the new direction is linearly proportional to $\Delta\varepsilon_0^2$, which is often proportional to the acoustic power at the frequency Ω in the small signal approximation. Usually Ω (in MHz or GHz) $<< \omega$, thus the small θ assumption used in Eq. (3.44) is justified.

Notes

1. RF is an abbreviation for radio frequency. It emphasizes that the applied electric field is a time varying field at microwave frequencies.
2. In principle, $\underline{F}$ can be any electric field including the electric field at the optical frequencies. However, the electric field of optical radiation is much smaller than the applied RF or DC electric field. Therefore the electro-optic effect refers usually to the change of index ellipsoid by the RF or DC electric field.

3. The values of r coefficients given here are for the 0.6328 μm optical wavelength and a RF electric field. Their values will vary slightly when the wavelength changes. However, the r coefficients are generally the same for all visible light and infrared wavelengths, they are independent of the frequency of the applied electric field, from DC to millimeter waves.

4. Typically semiconductor waveguides are fabricated by epitaxial growth of the core and cladding layers that are parallel to the substrate surface (see Section 1.3.2). In order to apply the RF electric field most effectively, the core layer is usually an i layer sandwiched between n and p type semiconductor layers, and a reverse biased voltage is applied to the p–i–n junction. A voltage is applied across the ground and the signal electrodes. Thus F is usually in the direction of the cut of the sample. The channel waveguide is often formed by etching a ridge.

5. Optical radiation at 1 μm wavelength has a photon energy of 1.24 eV. Since $\nu = c/\lambda$, and $\Delta\lambda/\lambda = \Delta\nu/\nu$, $\Delta\lambda_{det}$ of 0.1 μm at $\lambda \cong 1$μm would correspond to a $\Delta\nu_{det}$ of 50 meV.

6. Rect(u) is 1 for $-1 < u < 1$, and rect(u) is 0 for $u \geq 1$ and $u \leq -1$.

References

1. A. Yariv, *Optical Electronics in Modern Communications*, Oxford University Press (1997).
2. W. S. C. Chang, *Principles of Lasers and Optics*, Chapter 7, Cambridge University Press (2005).
3. G. P. Agrawal, Semiconductor laser amplifiers. In *Semiconductor Lasers*, ed. G. P. Agrawal, AIP Press (1995).
4. S. Shimoda and H. Ishio, *Optical Amplifiers and Their Applications*, John Wiley and Sons (1994).
5. D. P. Schinke, R. G. Smith, and A. R. Hartman, Photodetectors. In *Semiconductor Devices for Optical Communication*, 2nd edition, ed. H. Kessel, *Topics in Applied Physics*, vol. **39**, Springer-Verlag (1982).
6. P. K. L. Yu and Ming C. Wu, Photodiodes for high performance analog links, Chapter 8 in *RF Photonic Technology in Optical Fiber Links*, ed. W. S. C. Chang, Cambridge University Press (2002).
7. A. Yariv, *Quantum Electronics*, 3rd edition, John Wiley and Sons (1989).
8. L. A. Coldren and S. W. Corzine, *Diode Lasers and Photonic Integrated Circuits*, John Wiley and Sons (1995).
9. R. J. Keyes, ed., Optical and infrared detectors. In *Topics in Applied Physics*, Vol. **19**, Springer-Verlag (1980).
10. M. Born and E. Wolf, *Principles of Optics*, Section 14.2, Pergamon Press (1959).
11. A. Yariv, *Introduction to Optical Electronics*, Holt, Rinehart and Winston (1976).
12. H. S. Nalwa, T. Watanabe, and S. Miyata, Organic materials for second-order nonlinear optics. In *Nonlinear Optics of Organic Molecules and Polymers*, ed. H. S. Nalwa and S. Miyata, CRC Press (1997).
13. Y. M. Cai and A. K.-Y. Jen, Thermally stable poled polyquinoline thin film with very large electro-optic response. *Appl. Phys. Lett.*, **67** (1995) 299.
14. T. Van Eck, Polymer modulators for RF photonics, Chapter 7 in *RF Photonic Technology in Optical Fiber Links*, ed. W. S. C. Chang, Cambridge University Press (2002).
15. S. M. Sze, *Physics of Semiconductor Devices*, John Wiley and Sons (1981).
16. S. Wang, *Fundamentals of Semiconductor Theory and Device Physics*, Prentice Hall (1989).
17. B. G. Streetman, *Solid State Electronic Devices*, Prentice Hall (1995).
18. S. L. Chung, *Physics of Optoelectronic Devices*, John Wiley and Sons (1995).

19. W. Frantz, Einfluß eines elektrischen Felds auf eine optische Absorptionskante. *Z. Naturforschg*, **13a** (1958) 484.
20. L. V. Keldysh, The effect of a strong electric field on the optical properties of insulating crystals. *Sov. Phys., JETP*, **7** (1958) 788.
21. J. D. Dow and D. Redfield, Theory of exponential absorption edges in ionic and covalent solids. *Phys. Rev. Lett.*, **26** (1971) 762.
22. J. D. Dow and D. Redfield, Toward a unified theory of Urbach's rule and exponential absorption edges. *Phys. Rev. B*, **5** (1972) 594.
23. M. J. Sun, K. H. Nicholas, W. S. C. Chang, *et al.*, Gallium arsenide electro-absorption photodiode waveguide detectors. *Appl. Optics*, **17** (1978) 1568.
24. R. B. Welstand, High linearity modulation and detection in semiconductor electro-absorption waveguides. Section 2.2.2, Ph.D. thesis, University of California San Diego (1997).
25. Xiaobing Mei, InAsP/GaInP strain-compensated multiple quantum wells and their optical modulator applications. Ph.D. thesis, University of California San Diego (1997).
26. X. B. Mei, K. K. Loi, H. H. Wieder, W. S. C. Chang, and C. W. Tu, Strain compensated InAsP/GaInP multiple quantum wells for 1.3 μm waveguide modulators. *Appl. Phys. Lett.*, **68** (1996) 90.
27. D. A. B. Miller, D. S. Chemla, T. C. Damen, *et al.*, Electric field dependence of optical absorption near the bandgap of quantum well structures. *Phys. Rev. B*, **32** (1985) 1043.
28. D. A. B. Miller, J. S. Wiener, and D. S. Chemla, Electric field dependence of linear optical properties in quantum well structures: Waveguide electro-absorption and sum rules. *IEEE J. Quant. Elect.*, **QE-22** (1986) 816.
29. D. S. Chemla, D. A. B. Miller, P. W. Smith, A. C. Gossard, and W. Wiegmann, Room temperature excitonic non-linear absorption and refraction in GaAs/GaAlAs multiple quantum well structures. *IEEE J. Quant. Elect.*, **QE-20** (1984) 25.
30. T. L. Wood, J. Z. Pastalan, C. A. Burns, *et al.*, Electric field screening by photogenerated holes in multiple quantum wells: A new mechanism for absorption saturation. *Appl. Phys. Lett.*, **57** (1990) 1081.
31. A. Yariv, *Quantum Electronics*, 2nd edition, Appendix 1, John Wiley and Sons (1975).
32. K. K. Loi, Multiple-quantum-well waveguide modulators at 1.3 μm wavelength. Ph.D. thesis, University of California San Diego (1998).
33. D. A. Pinnow, Guidelines for the selection of acousto-optic materials. *IEEE J. Quant. Elect.*, **QE-6** (1970) 223.
34. Chen S. Tsai, ed., *Guided Wave Acousto-Optics*, Springer-Verlag (1990).
35. T. M. Reeder, Excitation of surface-acoustic waves by use of interdigital electrode transducers. In *Guided Wave Acousto-Optics*, ed. Chen S. Tsai, Springer-Verlag (1990).
36. R. W. Dixon, Photo elastic properties of selected materials and their relevance for applications to acoustic light modulators and scanners. *J. Appl. Phys.*, **38** (1967) 5149.

4   Time dependence, bandwidth, and electrical circuits

In order to create the electro-optical effects discussed in Chapter 3, a voltage is applied to the electrodes of the devices through electrical circuits to produce the electrical field. Most of the voltages that control the modulation, switching and signal processing functions are time varying, their frequency spectra range from MHz to tens of GHz. In analog applications, it is the frequency response of the device that is important whereas in digital applications, it is the time response of the device to a voltage (or current) pulse that is important. Pulse modulators are usually large signal devices. The time response of devices such as intensity modulation in a Mach–Zehnder or electro-absorption modulator is usually non-linear with respect to the magnitude of the applied voltage. Thus it is difficult to give a general discussion of the time response of electro-optical devices. However, pulses can be represented as a summation of their frequency components. Section 4.2.6 discusses the relation between frequency response and pulse propagation. Therefore, only the response of the devices to a time harmonic small voltage signal at different frequencies will be discussed in this book.

There are two major causes for frequency variation of the small signal response of electro-optical devices.

(1) The voltage across the device supplied by the electrical circuit is frequency dependent. For example, when the electrical source has a time harmonic variation, the fraction of the source voltage that appears across the device is frequency dependent. There is an electrical bandwidth of the voltage produced by the circuit driving the optoelectronic device.

(2) At high frequencies and for a given voltage applied to the input of the device, electrical signals propagate at microwave velocities on electrodes that function like transmission lines. The electrical field at any instant of time is not uniformly distributed over the length of the device. At the same time, optical guided waves propagate with a phase velocity of c/n_{eff} where c is the velocity of light in the free space. The modulation or switching effect is created through the electro-optical interaction of a traveling microwave with a traveling optical guided wave. There is a frequency dependence of the traveling wave interaction.

In addition to the frequency response, the signal voltage needs to be applied to the electrode from the RF source. It is important to do this efficiently. It is important to minimize the electrical power needed to drive the device. Properties of electrical circuits and propagation of electrical signals are reviewed in this chapter, followed by discussion of the various guided-wave devices and their time response in the chapters that follow.

4.1   Low frequency properties of electro-optical devices

4.1.1   Low frequency representation of devices

At low electrical frequencies or in shorter devices, the time for electrical and optical waves to propagate through the optoelectronic device is much shorter than the time period of RF signals. Under these circumstances, the instantaneous electric field variation seen by the optical guided wave has approximately a uniform time dependence throughout the device.[1] For example, at 100 MHz, the free space wavelength of the RF signal is 3 m. Within a small fraction of a period of the RF signal, the instantaneous spatial electrical field distribution in any optoelectronic device up to centimeters long is approximately the same as the electric field distribution at DC. Lumped electrical circuit elements such as resistance, capacitance, and inductance are used to characterize the electrical behavior (such as charging and discharging the electrode and current conduction and induction) of the device. Lumped circuit element representations of the device are then used in circuit analysis to determine the voltage applied to the device.

Clearly, any calculation of electro-optical effects must be based fundamentally on electromagnetic analysis. The lumped circuit elements such as capacitance, inductance and resistance are used only to simplify the representation of the device to the external circuits so that the voltage and current applied to the device can be found from circuit analysis. We still need to know from electromagnetic analysis how to represent quantitatively the electro-optical devices by lumped circuit elements and how to calculate the electric field.

Fundamentally, the microwave electric field $\underline{E}$ and magnetic field $\underline{H}$ in any device is calculated from electromagnetic theory. We know that the time averaged stored electric energy W_E and stored magnetic energy W_H for an electromagnetic field with $e^{j\omega t}$ time dependence in any volume (Vol) are

$$W_E = \frac{1}{4}\mathrm{Re}\int_{\mathrm{Vol}} \underline{E} \bullet \underline{D}^* \mathrm{d}v, \tag{4.1}$$

$$W_H = \frac{1}{4}\mathrm{Re}\int_{\mathrm{Vol}} \underline{H} \bullet \underline{B}^* \mathrm{d}v. \tag{4.2}$$

Here, Re is the real part of the integral, and Vol is the volume of the device. If a capacitance C and an inductance L are used to represent this device, its stored electric energy is $C|V|^2/4$, and its stored magnetic energy is $L|I|^2/4$. Therefore

$$C = 4W_E/V^2, \tag{4.3}$$

$$L = 4W_H/I^2, \tag{4.4}$$

where V is the voltage across the electrodes, and I is the current flowing in the electrodes. In more complex devices, where different parts of the device are represented by different lumped circuit elements, the Vol is just the volume of that portion of the device which the specific circuit element represents.

If there is time averaged power dissipation P_d due to the finite conductivity of the electrodes, then the resistance R representing that effect can be obtained from

$$R = 2P_d/|I|^2. \tag{4.5}$$

Please note that there are different ways in which power is dissipated; I in Eq. (4.5) is the current in the conductor, and P_d is the power dissipated in the conductor due to its finite conductivity. In the case of a leakage current across the electrodes, a shunt conductance G is used to represent time averaged power dissipation across the junction,

$$G = 2P_d/|V|^2. \tag{4.6}$$

For a microwave electric field that has time harmonic variation $e^{j\omega t}$,

$$P_d = \frac{\omega \varepsilon_0 \chi''}{2} \int_{\text{Vol}} \boldsymbol{E} \bullet \boldsymbol{E}^* dv. \tag{4.7}$$

A similar expression for P_d can be obtained in terms of the magnetic field $\boldsymbol{H}$.

In summary, from Eq. (4.1) to (4.7), the lumped elements, C, L and R (or G), representing the optoelectronic devices can be found from W_E, W_H and P_d which, in turn, can be measured or calculated from electromagnetic analysis.

For the electrodes of devices fabricated on insulators such as $LiNbO_3$, polymers or insulating semiconductors, the metal electrodes are considered as perfect conductors at low frequencies with no power dissipation. The inductance L is small enough so that its impedance, ωL, is commonly neglected at low frequencies. Their electrical characteristic is represented simply by a capacitance C. Occasionally, a shunt conductance G is used to represent power dissipation due to leakage current through the dielectrics. For semiconductor devices utilizing a reverse biased p–i–n junction to create the electric field, there may be contact resistance from the metal electrode to the i layer. We may then represent the p–i–n junction by a capacitance C (representing the portion of the device which is just the i layer) in series with a series contact resistance R_s (representing the portion of the device between the i layer and the metal electrode). If there is significant leakage current in the p–i–n junction, then there will be an additional junction conductance G in parallel with C. The major difference between the two cases lies not in the representation of the devices in electrical circuit analysis to determine V, but in calculating the electrical field for a given V on the electrode. For example, in the case of devices using p–i–n junctions, factors such as built-in-potential of the junction must be included in the calculation (see, for example, [1]).

4.1.2 Frequency variation of voltage and power delivered to devices

For most RF sources (or RF amplifiers), their output is represented by an ideal voltage source V_s in series with an internal resistance R_{source}. Figure 4.1 shows the electrical circuit of the source driving directly a modulator (or switch) in the lumped

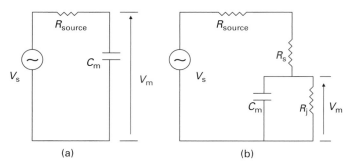

Fig. 4.1.   Electrical circuit of an electro-optical modulator driven directly by a voltage source. (a) Electro-optic modulators on insulators such as LiNbO$_3$ or polymer. (b) Electro-optic modulators using reverse biased p–i–n junction.

element representation, without any circuit element to match the modulator to the source. Figure 4.1(a) represents the case of a LiNbO$_3$ (or other insulating piezo-electric) modulator, while Fig. 4.1(b) depicts the case of a semiconductor modulator using a p–i–n junction to create the electric field. The voltage applied to the electro-optical device is V_m.

In the circuit configuration shown in Fig. 4.1(a), the impedance of the capacitance, $1/j\omega C_m$, is very large for small ω so that $V_m = V_s$ at very low frequencies. Note that V_m drops as ω is increased, and $|V_m| = V_s/\sqrt{2}$ when $\omega C_m R_{source} = 1$. In Fig. 4.1(b), since the leakage current through a reverse biased p–i–n junction is small, $R_j \gg R_{source}$. In addition the contact resistance R_s is much smaller than the junction resistance R_j. We obtain again $V_m \approx V_s$ at very low frequencies and $|V_m| \cong V_s/\sqrt{2}$ when $\omega C_m R_{source} = 1$. Therefore the low-pass 3 dB bandwidth[2] of this circuit configuration is $\omega = 1/C_m R_{source}$.

In the circuits shown in Fig. 4.1, all or most of the power provided by V_s is absorbed in the internal resistance R_{source}. In order to maximize the RF power delivered to the modulator, various circuit elements are used to match the modulator to the source. In the simplest matching scheme, a matching resistor R_{match} is usually placed in parallel with the modulator as shown in Fig. 4.2(a). At very low frequencies, the impedance of the capacitance is so large that all the current passes through the matching resistor. Therefore the power delivered to the modulator including the matching resistor is

$$\frac{V_s^2}{R_{match}} \left(\frac{1}{1 + \dfrac{R_{source}}{R_{match}}} \right)^2 .$$

A maximum amount of RF power, $V_s^2/4R_{match}$, is delivered when $R_{match} = R_{source}$. At higher frequencies, the RF power delivered to the modulator including the matching resistor is

$$\frac{V_s^2}{4R_{source}} \Big/ \left(1 + (R_{match}\omega C_m/2)^2 \right).$$

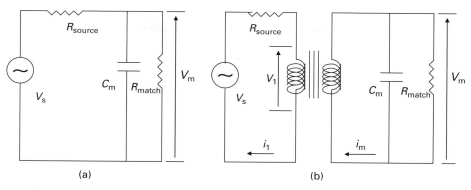

Fig. 4.2.   Matching of electro-optic modulators to the microwave source. (a) Modulator matched to the source by a simple matching resistor. (b) Modulator matched to the source by circuits represented as a transformer.

The RF power delivered is reduced by 3 dB when $\omega C_m R_{match} = 2$. Therefore, the low-pass bandwidth of this circuit configuration is $\omega = 2/R_{source} C_m$. Usually, $R_j \gg R_{match} \gg R_s$, therefore this result is also applicable approximately to the case of a p–i–n junction semiconductor modulator.

As another example of how the bandwidth is affected by the circuit configuration, consider the circuit shown in Fig. 4.2(b). In this case the impedance of the modulator is matched to the source by circuits represented as an equivalent ideal transformer. The ratio of the voltage and current transformed is $V_1/V_m = 1/N_m$ while $i_1 = N_m i_m$. Then the RF power delivered to the modulator including the matching resistor is

$$V_s^2 \left(\frac{R_{match}}{R_{source}^2 N_m^2} \right) \Bigg/ \left(1 + \frac{R_{match}}{R_{source} N_m^2} \right)^2 + (\omega C_m R_{match})^2.$$

At $R_{match} = N_m^2 R_{source}$, the maximum delivered RF power is

$$\frac{V_s^2}{4 R_{source}} \Bigg/ 1 + \left(\frac{N_m^2 \omega C_m R_{source}}{2} \right).$$

Clearly the 3 dB bandwidth of the maximum delivered power in this circuit is $\omega = 2/N_m^2 C_m R_{source}$. Similar results are obtained for p–i–n junction semiconductor modulators.

Other circuit configurations could be designed to give a resonance effect so that V_m is very large at a given frequency over a narrow bandwidth, or a band pass effect so that V_m is large over a specific bandwidth centered at a specific frequency. However, the basic problem is that C_m is fairly large for most devices. It is difficult to match a large C_m to the internal resistance of the source at high ω. In the three cases that we discussed above the basic bandwidth limitation is caused by $R_{source} C_m$. It is called the RC time-constant bandwidth limitation because, for any sudden change of V_s, the instantaneous voltages and currents will decay or increase exponentially with a decay constant in the order of $R_{source} C_m$.

4.2   High frequency properties of electro-optical devices

At high frequencies, the impedance of the inductance, ωL, which we neglected in Section 4.1 is now significant. The finite conductance of the electrodes may no longer be ignored. The voltage and current on the electrodes that provide the electric field for the electro-optical effects are now both a function of distance of propagation and time. We can no longer represent the device accurately by a single lumped circuit element of L, R or/and C. However, within a short distance Δz along the electrode, the transit time for electrical signals to propagate through dz is still much smaller than the time period of signal variation. We could still represent the Δz segment of the device by L, R, and C.

4.2.1   Representation of the electrodes as a transmission line

Figure 4.3 shows typical circuit elements, with the voltages and the currents of an incremental length of the electrode. Here Z_L is the series impedance per unit length. It often consists of an inductance L, representing the inductance of the electrodes, in series with a resistance R_c, representing the conductor loss of the electrodes, for the currents flowing along the electrodes. Note that $Z_L \Delta z$ is the series impedance between the input and the outputs of the segment, where $Z_L = R_c + j\omega L$ and Y_C is the parallel admittance per unit length. Note that $Y_C \Delta z$ is the impedance of the segment in parallel with the input and output. It often consists of a capacitance C, representing the displacement current due to charging and discharging of the electrodes across the dielectrics or the p–i–n junction, in parallel with a conductance G, representing the leakage current across the electrodes, and so $Y_C = j\omega C + \dfrac{1}{R_j}$. Then L, C, R_c and R_j can be calculated from $\underline{E}$ and $\underline{H}$ by Eq. (4.3) to (4.7), according to the physical configuration of the device. Notice that, in these equations, Vol now consists of a cylinder that includes the cross-section of the device and unit length along the electrodes. No matter what the circuit elements represented in Z_L and Y_C, we obtain

$$V(z + \Delta z) - V(z) = -I(z) Z_L \Delta z,$$

$$I(z + \Delta z) - I(z) = -V(z) Y_C \Delta z.$$

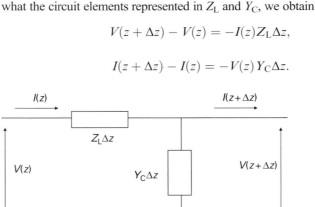

Fig. 4.3.   Impedance representation of a very short section of the electrodes at high frequencies.

As $\Delta z \to 0$, we obtain

$$\frac{\partial V}{\partial z} = -Z_L I \quad \text{and} \quad \frac{\partial I}{\partial z} = -Y_C V. \tag{4.8}$$

Equation (4.8) is the well known transmission line equation. It applies to the entire transmission line. For a transmission of any given length, solutions of Eq. (4.8) determine the V and I in the electrode. For an electrode and waveguide structure with Z_L and Y_C independent of z, the solution is

$$V(z, t) = V(z)e^{j\omega t} = \left[V^f e^{-\gamma z} + V^b e^{+\gamma z} \right] e^{j\omega t}, \tag{4.9}$$

$$I(z, t) = I(z)e^{j\omega t} = \left[I^f e^{-\gamma z} - I^b e^{+\gamma z} \right] e^{j\omega t} = \left[\frac{V^f}{Z_o} e^{-\gamma z} - \frac{V^b}{Z_o} e^{+\gamma z} \right] e^{j\omega t}, \tag{4.10}$$

where

$$\gamma = \sqrt{Z_L Y_C} = \frac{Z_L}{Z_o} = \alpha + j\beta. \tag{4.11}$$

Clearly the solution, $e^{-\gamma z + j\omega t} = e^{-\alpha z} e^{-j(\beta z - \omega t)}$, is a propagating wave in the $+z$ direction with phase velocity ω/β, wavelength $2\pi/\beta$, and amplitude attenuation, $\exp(-\alpha z)$. Similarly, the solution, $e^{+\gamma z + j\omega t} = e^{\alpha z} e^{j(\beta z + \omega t)}$, is a propagating wave in the $-z$ direction with phase velocity, ω/β, and amplitude attenuation, $\exp(\alpha z)$. Note that V^f and V^b are determined by the initial conditions at the input and output of the uniform transmission line. Also note that I^f is related to V^f by $I^f = V^f/Z_o$ while I^b is related to V^b by $I^b = V^b/Z_o$, where Z_o is known as the characteristic impedance of the transmission line, and α is the propagation loss coefficient, and β is the propagation constant for both the forward and backward waves. The quantity ω/β is the phase velocity.

When the electromagnetic field produced by the applied voltage is transverse electric and magnetic (TEM), the voltage V is related to the microwave E by

$$V = \int_{\text{ground electrode}}^{\text{signal electrode}} \underline{E} \bullet \underline{\mathbf{ds}}.$$

For other electrical waveguide configurations V is still proportional to E, and VI represents the electrical power transmitted through the transmission line [2]. *It is important to note that TEM electric field is assumed in most transmission line analysis in the literature.*

Electrodes in electro-optical devices are often designed in the form of popular microwave transmission lines such as the micro-strip line, the coplanar waveguide, etc. The Z_L and Y_C of these lines have been discussed in microwave literature [2]. A discussion of the field distribution and circuit representation of some transmission lines commonly used in optoelectronic devices is given by Chung [3]. There are also many computer simulation programs that can be used to calculate the transmission line field distribution and parameters from the configuration of the device [4]. Naturally, transmission line equations and the circuit parameters can also be derived directly from electromagnetic analysis [2].

4.2.2   Propagation of electrical voltages and currents

As an example to demonstrate the significance of initial conditions, let us consider a transmission of length l, from $z = -l$ to $z = 0$. Let it be terminated at $z = 0$ with an impedance Z_t. Then the initial condition at $z = 0$ is

$$V(z = 0) = Z_t I(z = 0). \tag{4.12}$$

Therefore,

$$V(z) = \frac{V(z = 0)}{2} [(Z_t + Z_o)e^{-\gamma z} + (Z_t - Z_o)e^{+\gamma z}], \tag{4.13}$$

$$I(z) = \frac{V(z = 0)}{2Z_o} [(Z_t + Z_o)e^{-\gamma z} - (Z_t - Z_o)e^{+\gamma z}]. \tag{4.14}$$

When $Z_t = Z_o$, there is no backward propagating wave, i.e. no reflection at $z = 0$. The transmission line is said to be match terminated. In a matched transmission line, $V(z) = Z_t I(z)$ at all positions of z. On the other hand, for $Z_t \gg Z_o$ or $Z_t \ll Z_o$, the forward and backward propagation waves will have nearly equal amplitudes, $V^f = (I(z = 0)/2)Z_t = V^b$ or $V^f = (I(z = 0)/2)Z_o = -V^b$. In practice, Z_t may be mismatched somewhat from Z_o. It will then have partially reflected waves. Usually a RF source with internal impedance Z_{source} is also connected to the transmission line on the input end at $z = -l$. When Z_{source} is matched to Z_o, there will be no reflections at the source end. When $Z_{source} \neq Z_o \neq Z_t$ there are multiple forward and backward propagating waves.

For an ideal lossless transmission line discussed in most textbooks, $Z_L = j\omega L$ and $Y_C = j\omega C$. Then $\alpha = 0$ and Z_o is a real constant independent of ω, $Z_o = \sqrt{L/C}$ and $\omega/\beta =$ phase velocity $= 1/\sqrt{LC}$. In a matched lossless ideal transmission line, $V(z=0) = Z_o I(z=0)$. The electro-optical device can now be represented to the driving circuit as a resistor at the input instead of the capacitor shown in Fig. 4.1(a). *The significance of this substitution is that when $Z_o = Z_t = R_{source}$, the matching of the device impedance to the source impedance is good at all frequencies. There is no RC time constant limitation of the bandwidth in the circuit.*

In the above discussions, we have used circuit elements L and R_c in Z_L, and C and R_j in Y_C. This is the most common case. There may be other elements such as the contact resistance shown in Fig. 4.1(b). However, Eq. (4.8) to (4.14) are also valid for more complex Z_L and Y_C. When Z_L and Y_C vary with z, the transmission line can be considered as segments of transmission that have a constant Z_L and Y_C within each segment. The initial conditions at the junctions of each segment determine the relation between the forward and backward waves. Often, there may also be partial reflections caused by discontinuities of the characteristic impedances of the different segments of the transmission line or by other devices connected to the transmission line at different locations. These discontinuities will produce a complex frequency response of the transmission line.

If we define Γ as the reflection coefficient of V^b from V^f, i.e.

$$\Gamma_r = V^b / V^f, \tag{4.15}$$

then Eq. (4.9) and (4.10) can be written for any $V(z = 0)/I(z = 0)$ as

$$V(z) = V^f(z = 0)[e^{-\gamma z} + \Gamma_r(z = 0)e^{+\gamma z}], \tag{4.16}$$

$$I(z) = \frac{V^f(z = 0)}{Z_o}[e^{-\gamma z} - \Gamma_r(z = 0)e^{+\gamma z}]. \tag{4.17}$$

The input impedance at $z = -l$, Z_{in}, can be expressed as

$$Z_{in} = \frac{V(z = -l)}{I(z = -l)} = Z_o \frac{1 + \Gamma_r(z = 0)e^{-2\gamma l}}{1 - \Gamma_r(z = 0)e^{-2\gamma l}}. \tag{4.18}$$

4.2.3   The Smith chart

The Smith chart is a plot of the complex reflection coefficient in both rectangular and polar coordinates.

The Z_{in} in Eq. (4.18) is a function of the reflection coefficient, $\Gamma_r(z = 0)e^{-2\gamma l}$. The reflection coefficient, $\Gamma_r(z = 0)e^{-2\gamma l}$, is a complex quantity. Mathematically, any complex quantity can be expressed either as a quantity with a real part and an imaginary part in rectangular coordinates or as a vector in polar coordinates with a magnitude and a phase angle, thus

$$\Gamma_{re} + j\Gamma_{im} = \Gamma_r(z = 0)e^{-2\gamma l} = |\Gamma_r(z = 0)|e^{j\theta}. \tag{4.19}$$

Since $|\Gamma_r(z = 0)| \leq 1$ and $\alpha > 0$ for all passive components, any $\Gamma_r(z = 0)e^{-2(\alpha + j\beta)l}$ can be plotted as a vector (or point) in the polar coordinate system within a circle of unity. Let the horizontal axis be the real axis and the vertical axis be the imaginary axis. The magnitude is plotted as a radius from the center of the circle, up to unity. The phase angle zero is represented as the horizontal radius to the right. The phase angle $-2j\beta l$ can be expressed as a rotation of angle from the horizontal radius in the polar plot. Since $e^{j2\pi} = 1$, only $-\pi \leq (2j\beta l - 2m\pi) \leq \pi$ needs to be plotted. Here m is an integer. As βl increases, the phase angle rotates clockwise from the horizontal radius. Vice versa, the phase angle rotates counter clockwise when βl decreases. The Smith chart is basically a plot of $\Gamma_r(z = 0)e^{-2(\alpha + j\beta)l}$ simultaneously in rectangular and polar coordinates. Figure 4.4 shows the Smith chart. Any $\Gamma_r(z = 0)e^{-2(\alpha + j\beta)l}$ appears as a point within the circle of unity centered at $\Gamma_{re} = 0$ and $\Gamma_{im} = 0$.

A lot of information about the transmission line can be learned when $\Gamma_r(z = 0)e^{-2(\alpha + j\beta)l}$ is plotted as βl is varied. For example, when the microwave frequency is varied from low to high, β is increased. For transmission lines described by Z_L

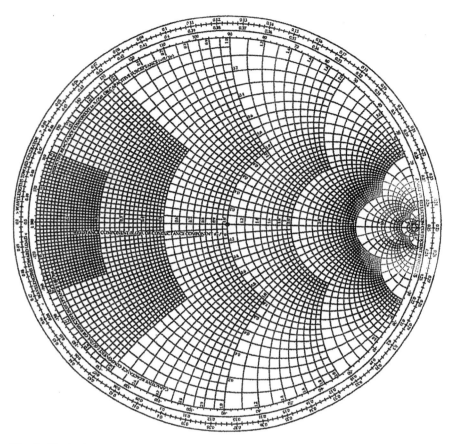

Fig. 4.4.   The Smith chart.

and Y_C discussed in Section 4.2.1, β is given by Eq. (4.11). If α is independent of frequency, the magnitude of the vector does not change. Only the rotation angle changes with frequency. If the α is increased at higher frequencies, as many microwave transmission lines do, then the magnitude of the vector decreases at higher frequencies. Vice versa, if $\Gamma_r(z = 0)e^{-2(\alpha+j\beta)l}$ is measured experimentally for a given l as the frequency is varied, then plotting it in the polar form will let us verify or determine Z_L and Y_C.

The Smith chart also allows us to determine Z_{in}/Z_o directly and graphically from the plot of $\Gamma_r(z = 0)e^{-2(\alpha+j\beta)l}$. From Eq. (4.18), it is clear that the Z_{in} normalized with respect to Z_o is

$$\frac{Z_{in}}{Z_o} = r_z + jx_z = \frac{1 + (\Gamma_{re} + j\Gamma_{im})}{1 - (\Gamma_{re} + j\Gamma_{im})}. \tag{4.20}$$

Therefore

$$r_z = \frac{1 - \Gamma_{re}^2 - \Gamma_{im}^2}{(1 - \Gamma_{re})^2 + \Gamma_{im}^2}, \tag{4.21}$$

$$x_z = \frac{2\Gamma_{\text{im}}}{(1 - \Gamma_{\text{re}})^2 + \Gamma_{\text{im}}^2}. \tag{4.22}$$

Rearranging Eq. (4.21) and (4.22), we obtain

$$\left(\Gamma_{\text{re}} - \frac{r_z}{1 + r_z}\right)^2 + \Gamma_{\text{im}}^2 = \left(\frac{1}{1 + r_z}\right)^2, \tag{4.23}$$

$$(\Gamma_{\text{re}} - 1)^2 + \left(\Gamma_{\text{im}} - \frac{1}{x_z}\right)^2 = \left(\frac{1}{x_z}\right)^2. \tag{4.24}$$

Equations (4.23) and (4.24) represent two families of circles in the Γ_{re} and Γ_{im} plane. For each constant r_z value, Eq. (4.23) describes a set of circles with center at $\Gamma_{\text{re}} = r_z/(1 + r_z)$ and $\Gamma_{\text{im}} = 0$ and radius $1/(1 + r_z)$. These are shown as the set of resistance circles in Fig. 4.4. All resistance circles pass through $\Gamma_{\text{re}} = 1$ and $\Gamma_{\text{im}} = 0$. For each constant x_z value, Eq. (4.24) describes a set of circles with center at $\Gamma_{\text{re}} = 1$ and $\Gamma_{\text{im}} = 1/x_z$ and radius $1/x_z$. These are shown as the set of reactance circles in Fig. 4.4. All reactance circles also pass through $\Gamma_{\text{re}} = 1$ and $\Gamma_{\text{im}} = 0$. However, their centers are always at $\Gamma_{\text{re}} = 1$ on the real axis and $\Gamma_{\text{im}} = 1/x_z$ on the imaginary axis, which varies according to the value of x_z. For a given transmission line, as soon as the "$\Gamma_{\text{r}}(z = 0)e^{-2(\alpha + j\beta)l}$" is plotted, the r_z and x_z of the input impedance Z_{in} can be read directly from these resistance and reactance circles.

4.2.4 Characterizing the electrodes as electrical transmission lines and circuit analysis

Determination of V (or V^f and V^b), I (or I^f and I^b), Z_o, α and β, is very important in order to analyze and to understand the performance of electro-optical devices at high frequencies. There is a large amount of information and analytical techniques in microwaves that can be used to determine V and I. More importantly, α, β and Z_o, determined theoretically may not be accurate under many circumstances. Values of V and I at the input or the output may be the only quantities that can be experimentally measured. It is often necessary to use the experimental results to determine or to verify the values of α, β and Z_o (i.e. the Z_L and Y_C). The values of the circuit elements in Z_L and Y_C are adjusted numerically so that the calculated results will fit the measured results.

The analysis of the total circuit, including the driving and termination circuit and the transmission line using the experimentally verified values of α, β, and Z_o, will determine the voltage variation along the electrodes. Usually, the electrodes are also connected to the external circuits via transitional transmission lines. Just like the analysis in low frequency circuits, we need to know how to calculate or measure the propagating waves on the electrodes via transmission line analysis as a part of the total circuit. For these reasons some electrical transmission line network notations and analyses will be reviewed here.

4.2.4.1 The ABCD transmission matrix representation

Typically, the transmission line representing an electro-optical device is connected electrically to a driving circuit at the input and to a terminal impedance (or circuit) at

Fig. 4.5.   Two port network representation of the transmission line electrode.

the output. Therefore the transmission line is represented conventionally as a two port electrical network, as shown in Fig. 4.5. Here V_1 and I_1 are the voltage and current at the input, and V_2 and I_2 are the voltage and current at the output. For a given terminal and source circuit configuration, V_1, V_2, I_1, and I_2 are solved by network analysis. They are the initial conditions that will determine V^f and V^b (or I^f and I^b). For a transmission line that has linear and reciprocal elements

$$V_1 = AV_2 + BI_2$$
$$I_1 = CV_2 + DI_2. \tag{4.25}$$

Or, in the matrix notation,

$$\begin{vmatrix} V_1 \\ I_1 \end{vmatrix} = \begin{vmatrix} A & B \\ C & D \end{vmatrix} \begin{vmatrix} V_2 \\ I_2 \end{vmatrix}, \tag{4.26}$$

where

$$AD - BC = 1. \tag{4.27}$$

Note that when the directions of current and voltage are used in circuit analysis to indicate the relative phase of time harmonic variations, I_1 and I_2 are defined with directions opposite to each other in Fig. 4.5.

4.2.4.2   The impedance matrix representation

Relations between V_1, V_2, I_1, and I_2 could also be expressed as

$$V_1 = Z_{11}I_1 + Z_{12}I_2$$
$$V_2 = Z_{21}I_1 + Z_{22}I_2. \tag{4.28}$$

Or, in the matrix notation,

$$\begin{vmatrix} V_1 \\ V_2 \end{vmatrix} = \begin{vmatrix} Z_{11} & Z_{12} \\ Z_{21} & Z_{22} \end{vmatrix} \begin{vmatrix} I_1 \\ I_2 \end{vmatrix}, \tag{4.29}$$

where $Z_{12} = Z_{21}$. This is known as the impedance matrix representation. Clearly there is also a less frequently used admittance matrix representation expressing I_1 and I_2 in terms of V_1 and V_2. Elements of the impedance matrix are related to elements of the transmission matrix by

$$Z_{11} = A/C, \quad Z_{22} = D/C, \quad Z_{12} = Z_{21} = 1/C, \tag{4.30}$$

or

$$A = Z_{11}/Z_{21}, \quad B = (Z_{11}Z_{22} - Z_{12}Z_{21})/Z_{21},$$
$$C = 1/Z_{21}, \quad D = Z_{22}/Z_{21}. \tag{4.31}$$

4.2.4.3 The scattering matrix representation

In microwave measurements, one usually measures the magnitude and phase of a forward or a backward propagating wave at a given port by a vector network analyzer. Therefore it is convenient to represent the inputs and outputs to the physical transmission line by the amplitude and phase of its forward and backward waves at port 1 and port 2:

$$V_1 = V_1^+ + V_1^-, \quad V_2 = V_2^+ + V_2^-. \tag{4.32}$$

Notice that, in terms of the z direction used in Eq. (4.8) to (4.19), V_1^+ and I_1^+ are waves in the $+z$ direction, while V_2^+ and I_2^+ are waves in the $-z$ direction. Similarly V_2^- and I_2^- are in the $+z$ direction, while V_1^- and I_1^- are in the $-z$ direction. Since $I^f = V^f/Z_o$ and $I^b = V^b/Z_o$,

$$I_1 = I_1^+ - I_1^- = \frac{V_1^+}{Z_o} - \frac{V_1^-}{Z_o}, \tag{4.33}$$

$$I_2 = I_2^+ - I_2^- = \frac{V_2^+}{Z_o} - \frac{V_2^-}{Z_o}. \tag{4.34}$$

Conversely, V_1^+, V_1^-, I_2^+ and I_2^- (or V_1^+, V_1^-, V_2^+, and V_2^-) can also be expressed in terms of linear combinations of V_1, V_2, I_1 and I_2.

Knowing the relation between V_1, V_2, I_1, and I_2 is equivalent to knowing the relation between V_1^+, V_2^+, V_1^- and V_2^-. Therefore, just like the transmission and the impedance matrix representation, the backward waves at ports 1 and 2 are linearly related to the forward waves at ports 1 and 2 by a matrix, called the scattering matrix,

$$\begin{vmatrix} V_1^- \\ V_2^- \end{vmatrix} = \begin{vmatrix} S_{11} & S_{12} \\ S_{21} & S_{22} \end{vmatrix} \begin{vmatrix} V_1^+ \\ V_2^+ \end{vmatrix}. \tag{4.35}$$

The elements of the scattering matrix are related directly to the elements of the impedance matrix and the ABCD transmission matrix. For example

$$\frac{Z_{11}}{Z_o} = \frac{(1 + S_{11})(1 - S_{22}) + S_{21}S_{12}}{(1 - S_{11})(1 - S_{22}) - S_{21}S_{12}}, \tag{4.36}$$

$$\frac{Z_{22}}{Z_o} = \frac{(1 - S_{11})(1 + S_{22}) + S_{21}S_{12}}{(1 - S_{11})(1 - S_{22}) - S_{21}S_{12}}, \tag{4.37}$$

$$\frac{Z_{12}}{Z_o} = \frac{2S_{12}}{(1 - S_{11})(1 - S_{22}) - S_{21}S_{12}} = \frac{Z_{21}}{Z_o}. \tag{4.38}$$

The element S_{12} is obtained (or measured) from the reflected wave at port 1, V_1^-, when the transmission line is driven by a forward propagating wave V_2^+ at port 2 while V_1^+ is zero. This means that port 1 should be terminated with Z_o so that V_1^- will not produce any forward wave (i.e. $V_1^+ = 0$). Similarly, S_{11} is determined (or measured) from V_1^- when the transmission line is driven by V_1^+ with $V_2^+ = 0$ (i.e. with Z_o termination at port 2). Note that S_{22} is determined from V_2^- when the transmission line is driven by V_2^+ with $V_1^+ = 0$. In short,

$$S_{ij} = \left. \frac{V_i^-}{V_j^+} \right|_{V_m^+=0 \text{ for } m \neq j}. \tag{4.39}$$

Consider now the S_{21} of the transmission line described in Eq. (4.9) to (4.11). When a forward wave V_1^+ is incident on port 1 which is located at $z = -l$,

$$V_1^+ = V^f e^{\gamma l},$$

and V^b is zero because the transmission line is matched at $z = 0$, i.e. port 2. Therefore, at port 2

$$V_2^- = V^f \quad \text{and} \quad V_2^+ = 0.$$

In other words,

$$S_{21} = e^{-\gamma l}. \tag{4.40}$$

Or alternatively, S_{21} is just the reflection coefficient $\Gamma_r(z=0)e^{-2(\alpha+j\beta)l}$ with $\Gamma_r(z=0) = 1$. It can be plotted on the Smith chart. The measurement of S_{21} for a given l plotted on the Smith chart as ω is swept from low to high frequencies tells us about α and β (indirectly Z_L and Y_C through curve fitting) and the normalized Z_{in} can be read from the resistance and reactance circles directly for $\Gamma_r(z=0) = 1$. Note that $S_{21} = S_{12}$, and Eq. (4.39) implies also that $S_{11} = S_{22} = 0$.

4.2.5 Impedance matching and bandwidth

In practical applications, the electro-optical device is often connected to the microwave supply at port 1 through a transitional transmission line and terminated by an impedance Z_t at port 2. The microwave supply, such as the output from an amplifier, is represented as a microwave source in series with a source impedance R_{source}. The electrical representation of the supply is the same as the source circuit shown in Fig. 4.1. It should be clear from the preceding discussions that the goal of the design of the transitional transmission line and the electrodes for high frequency electro-optic devices is to have low loss transmission lines (for both the transitional transmission and the electrode) with $Z_o = 50$ ohms and $Z_t = 50$ ohms. Then the electro-optic device is matched to the microwave source impedance at all frequencies. In that case there are no reflected waves on the transmission line.[3] The frequency response (or bandwidth) of the device will be determined by the interaction of a forward traveling microwave with a forward traveling optical guided wave. Note that the attenuation of either the microwave or the optical wave will limit also the effective length of the traveling wave interactions. Since the attenuation of microwave transmission line increases at higher frequency, these two factors impose further bandwidth limitations on the device response at higher frequencies. For example, a great deal of research effort has been expended to obtain specific shapes and separation of electrodes to achieve the goal of low loss 50 ohms characteristic impedance for electrodes in LiNbO$_3$ modulators [5]. In practice, 50 ohms of Z_o is hard to get, one usually settles for a design that yields a lower Z_o and matches it with an appropriate Z_t and an impedance transformer to the source so that multiple reflections will not limit the bandwidth of the device.

4.2.6 Transient response

Despite best efforts to match the transmission lines to the source, there may be reflections caused by impedance mismatches or discontinuities at various points in the transmission lines caused by bending, transmission line mismatch or material discontinuity in the circuit which includes the source, the device, the termination and the transitional transmission line. In the measurement of the various parameters such as S_{ij} or Z_{ij} of the total circuit as a function of frequency, these discontinuities will yield unexpected bumps in the frequency variation. It is difficult to separate the problems from just the frequency measurements. A pulsed input from the RF source is then applied to the circuit, and the time behavior of the reflected pulse from the circuit including the device is monitored. Knowing the velocity, the time lapse between the transmitted and the reflected pulse then allows us to determine the position of the discontinuity in the total circuit that caused the reflection.

4.2.7 Pulse propagation and frequency response

In order to relate the frequency response to the pulse response, let $V(z,t)$ be a pulse of the signal voltage at center frequency $\overline{\omega}$,

$$V(z,t) = A(z,t)e^{-j(\overline{\beta}z - \overline{\omega}t)},$$

where $A(z,t)$ is the pulse envelope, $\overline{\beta}$ is the averaged propagation wave number, $\overline{\beta} = \overline{n}_{\text{eff}}k$, and $\overline{\omega}$ is the center frequency. In terms of Fourier Transform, it means that there is a group of CW signal voltage at ω centered about $\overline{\omega}$ such that

$$V(z,t) = A(z,t)e^{-j(\overline{\beta}z - \overline{\omega}t)} = \int a(\omega,z)e^{-j(\beta z - \omega t)}d\omega, \tag{4.41}$$

where $a(\omega)$ is the Fourier amplitude of the spectra component at ω. Note that β is obviously a function of ω, $a(\omega) \neq 0$ only when $\overline{\omega} - \Delta\omega < \omega < \overline{\omega} + \Delta\omega$, and $\Delta\omega$ is the spectral width. Usually $\Delta\omega/\Delta\overline{\omega} \ll 1$. It is well known that the smaller the $\Delta\omega$, the wider is the pulse duration.

Utilizing a Taylor series approximation to express the relation between $\beta(\omega)$ and $\overline{\beta}$, Eq. (4.41) can be rewritten as

$$A(z,t) = \int a(\omega,z)e^{j\left[(\omega-\overline{\omega})t-(\beta-\overline{\beta})z\right]}\,d\omega = \int a(\omega,z)e^{j\left[(\omega-\overline{\omega})\left(t-\frac{d\beta}{d\omega}\big|_{\overline{\beta}}z\right)\right]}\,d\omega.$$

Hence the velocity of advance of a definite value of A such as the maximum of A is given by the group velocity

$$v_g = \left(\frac{d\omega}{d\beta}\right)\Bigg|_{\overline{\beta}}. \tag{4.42}$$

From the preceding discussions, it is clear that the transmission line analysis will only give us the amplitude and phase of the forward and backward propagating waves of each spectral component, i.e. the $a(\omega)$ for each group of forward and backward waves. When there are forward and backward pulses, each individual group of $a(\omega)$ will travel with the group velocity given in Eq. (4.42). The group velocity will only give an estimation of the response expected from a pulse centered at $\overline{\omega}$. There is broadening or sharpening of the pulse as it propagates in the z direction. In an ideal device, where the electrodes are matched to the source and have proper termination, there will be just the forward propagating pulse.

4.3   Microwave electric field distribution and the electro-optical effects

For a given voltage on the electrode at a z location, it creates an electric field distribution. The electric field pattern in the xy plane is the same at different positions of z, but its amplitude is proportional to the voltage. Part of the electric field that overlaps the electro-optical active medium creates the electro-optical effects, including the linear electro-optic effect, the electro-absorption effect and the electro-refraction effect.

Electromagnetic solution of the electrodes fabricated on specific material structures is required to give the electric field distribution. For all device structures there is a linear relation between the microwave electric field and the voltage on the electrode. Electrode structures with two or more electrodes have TEM (transverse electric magnetic) modes, meaning $E_z = H_z = 0$. In this case, within each region such as core, cladding, etc., we have

$$\frac{\partial E_y}{\partial x} - \frac{\partial E_x}{\partial y} = 0 \quad \text{and} \quad \frac{\partial H_y}{\partial x} - \frac{\partial H_x}{\partial y} = 0. \tag{4.43}$$

The E_x, E_y, H_x and H_y are related by

$$E_x = \sqrt{\frac{\mu}{\varepsilon}}H_y \quad \text{and} \quad E_y = -\sqrt{\frac{\mu}{\varepsilon}}H_x. \tag{4.44}$$

Differentiating E_x and E_y in Eq. (4.43) with respect to x and to y once more and using Eq. (4.44), the Maxwell's equations simplify into a Laplace Equation (see [2] for derivation)

$$\left(\frac{\partial^2}{\partial x^2} + \frac{\partial^2}{\partial y^2}\right)\underline{E_t} = 0. \tag{4.45}$$

Here E_t denotes electric field in the transverse xy plane, $\underline{E_t} = E_x \underline{i_x} + E_y \underline{i_y}$.

A simple example of Eq. (4.45) is to calculate the electric field in a parallel plate capacitor with two parallel plate electrodes with infinite conductivity separated by a dielectric with distance of separation d and permittivity ε. In that case, for a voltage V

$$E_x = V/d. \tag{4.46}$$

For a parallel capacitor with finite electrode area A, we usually neglect its fringe electric field distribution near the edge of the electrode. Then its capacitance is obtained from Eq. (4.3) as

$$C = \varepsilon \frac{A}{d}. \tag{4.47}$$

Because of its simplicity the electric field and the capacitance of a parallel plate capacitor are used commonly to give a first estimation of the electric field of various electrode configurations.

Various electrostatic solutions of the Laplace Equations such as the scalar potential, Green's function and conformal mapping can be used to calculate the microwave electric field. For example the electric field, the Z_o, the effective index (i.e. β) and the attenuation of micro-strip electrodes similar to those shown in Fig. 1.9 are discussed in microwave textbooks [2]. The electric field of the thin symmetric and asymmetric coplanar electrode has been calculated by Ramer [6]. The microwave impedance, the effective index and the attenuation of a coplanar waveguide and asymmetric coplanar strip electrode on $LiNbO_3$ have been calculated by Chung and Chung et al. [3, 7]. These electrode configurations are similar to those shown in Fig. 3.2. Commercial software such as *ANSOFT* is used to calculate the transverse electric field distribution, Z_o, α, and β, for complex structures such as thick electrodes on a material structure which may contain a reverse biased p–i–n junction. Note that the electric field distribution for a given electrode configuration will vary slowly as ω is changed.

It is important to note that there are higher order modes which are not TEM modes. Since TEM modes give usually the lowest attenuation and the highest Z_o, they are the microwave modes which are intended to be used to create the electro-optical effect. Nevertheless, higher order modes could be excited in practice and they may affect significantly the β, the α and the Z_o of the transmission line representing the electrodes.

Only the microwave electric field that overlaps the electro-optical active medium will produce the electro-optical effect. This electric field will create a $\Delta\chi$, or $\Delta\varepsilon = \varepsilon_o \Delta\chi = \varepsilon_o(\Delta\chi' - j\Delta\chi'')$ by the linear electro-optic effect, the electro-absorption effect or the electro-refraction effect in the active medium. If this electric field is approximately uniform in spatial distribution, then the effect of $\Delta\chi$ on the propagation of the guided-wave mode can be obtained by substituting ε by $(\chi_o + \Delta\chi)\varepsilon_o$ for the active medium in calculating the n_{eff} and attenuation for the guided-wave mode. When the electric field is non-uniform, $\Delta\chi$ will be non-uniform. The change in the propagation constant can be obtained from substituting $\Delta\varepsilon$ into Eq. (2.6) using the perturbation analysis.

4.4   Traveling wave interactions

Let there be an optical nth guided wave, $Ae^{-\frac{a_{n,o}}{2}z}e^{-jn_{\text{eff},n}\beta_o z}e^{j\omega t}$, modulated by a traveling wave microwave electric field, $F_{\text{RF}}e^{-\alpha_{\text{RF}}z}e^{-j\beta_{\text{RF}}z}e^{j\omega t}$, where α_{RF} and β_{RF} of the microwave transmission line are given in Eq. (4.11). Note that F_{RF} is proportional to the microwave voltage V_{RF}, and β_{RF} is equal to $n_m\beta_o$, where n_m is the microwave effective index. The microwave creates a $\Delta\varepsilon$ similar to that discussed in Section 3.5, except that $\Delta\varepsilon$ is now time dependent. Note that $\alpha_{n,o}$ is the attenuation coefficient of the intensity of the nth order optical guided wave in the absence of RF field, and β_o is ω/c, the free space propagation wave number for both optical and microwave waves. Let us assume that both the optical waveguide and the microwave transmission line are matched. There is no reflected wave in the $-z$ direction.

The perturbation analysis, given in Eq. (2.1) to (2.6), was derived for a $\Delta\varepsilon$ independent of time. At any specific instant of time, Eq. (3.37) and (3.40) can still be applied to the optical guided wave for an instantaneous $\Delta\varepsilon$, over a small distance dz. The value of $\Delta\varepsilon$ changes as the optical guided wave propagates in time and distance. For the forward traveling wave microwave and the nth guided-wave mode assumed here, the instantaneous $\Delta\varepsilon$ seen by the guided wave as it propagates is

$$\Delta\varepsilon = \Delta\varepsilon_{\text{max}}e^{-\alpha_{\text{RF}}z}\cos(\omega t - \delta n\beta_o z), \quad \text{with} \quad \delta n = n_m - n_{\text{eff},n}. \tag{4.48}$$

Here, the microwave is assumed to be launched at $z = 0$, $\Delta\varepsilon_{\text{max}}$ is the maximum $\Delta\varepsilon$ at $z = 0$, and $\delta n\beta_o$ accounts for the difference in the propagation velocity of microwaves and optical waves. According to Eq. (3.33) and (3.34), $\Delta\varepsilon_{\text{max}}$ is complex,

$$\Delta\varepsilon_{\text{max}} = 2n\Delta n_{\text{max}}\varepsilon_o - j\Delta\chi''_{\text{max}}\varepsilon_o. \tag{4.49}$$

Let us now follow an optical guided wave from $z = 0$. Within each short section dz, the electro-optic Δn causes a phase shift $\Delta\beta$dz. Therefore, over a propagation distance L, the total phase shift created by electro-optic effects without electro-absorption is

$$\Delta\phi = \int_0^L \Delta\beta \, dz = \frac{\omega n\varepsilon_o}{2} \int_{\text{active region}} \Delta n_{\text{max}}(x,y)\left(\underline{e}_n \bullet \underline{e}_n^*\right) ds$$

$$\int_0^L e^{-\alpha_{\text{RF}}z}\cos(\omega t - \delta n\beta_o z) \, dz. \tag{4.50}$$

This is the result for a traveling wave interaction that corresponds to Eq. (3.37) for constant Δn. Let

$$AL\cos(\omega t - \xi) = \int_0^L e^{-\alpha_{\text{RF}}z}\cos(\omega t - \delta n\beta_o z) \, dz. \tag{4.51}$$

After direct integration, we obtain

$$A^2 L^2 = \frac{1}{(\alpha_{RF}^2 + \delta^2)} \left[\left(1 + e^{-2\alpha_{RF}L}\right) - 2\cos\delta n\beta_o L e^{-\alpha_{RF}L} \right], \tag{4.52}$$

$$\tan\xi = \frac{(-\alpha_{RF}\sin\delta n\beta_o L - \delta n\beta_o \cos\delta n\beta_o L)e^{-\alpha_{RF}L} + \delta n\beta_o}{(-\alpha_{RF}\cos\delta n\beta_o L + \delta n\beta_o \sin\delta n\beta_o L)e^{-\alpha_{RF}L} + \alpha_{RF}}. \tag{4.53}$$

Therefore,

$$\Delta\phi = \frac{\omega n\varepsilon_o}{2} \int\limits_{\text{active region}} \Delta n_{\max}(x,y)(\underline{e_n} \bullet \underline{e_n^*})\mathrm{d}s\, AL\cos(\omega t - \xi). \tag{4.54}$$

Similarly, the total attenuation of the amplitude of an optical guided wave, created by electro-absorption without electro-optic Δn, is

$$\exp\left(-\int\limits_0^L \frac{\Delta\alpha_n}{2}\mathrm{d}z\right) = \exp\left[-\frac{\omega}{4}\varepsilon_o \int\limits_{\text{active region}} \Delta\chi_{\max}''(\underline{e_n} \bullet \underline{e_n^*})\mathrm{d}s\right.$$

$$\times \int\limits_0^L e^{-\alpha_{RF}z}\cos(\omega t - \delta n\beta_o z)\mathrm{d}z\Big]$$

$$= \exp\left[-\frac{\omega}{4}\varepsilon_o \int\limits_{\text{active region}} \Delta\chi_{\max}''(\underline{e_n} \bullet \underline{e_n^*})\mathrm{d}s\, AL\cos(\omega t - \xi)\right]. \tag{4.55}$$

This is the result for traveling wave electro-absorption interaction, corresponding to Eq. (3.40) for the constant $\Delta\chi''$.

The factor $A\cos(\omega t - \xi)$ shows the effect of traveling wave interaction. Note that when the velocity of the optical guided wave matches that of the microwave, i.e. $\delta n = 0$, $A = 1$ and $\xi = 0$. This is the ideal case for traveling wave interaction where the electro-optical effect is identical for any modulation frequency ω. The electro-optical susceptibility does not impose a bandwidth limitation. As δ increases, "A" decreases.

Notes

1. It is important to note that "low frequency" refers only to the case when the electrical wavelength is much longer than the device length. For short devices only hundreds of micrometers long, the "low frequency" electrical representation may still be applicable even at low GHz frequencies.
2. Since power is proportional to V^2, the power delivered to the modulator is reduced by 3 dB when V is reduced by $1/\sqrt{2}$.
3. The transitional transmission line is often tapered or curved gradually so that the electrode transmission line can be matched to the source without reflection, conforming to the physical configuration of the devices.

References

1. R. B. Welstand, High linearity modulation and detection in semiconductor electroabsorption waveguides. Chapter 2, Ph.D. thesis, University of California San Diego (1997).
2. D. M. Pozar, *Microwave Engineering*, John Wiley and Sons (2005).
3. Haeyang Chung, Optimization of microwave frequency traveling-wave LiNbO$_3$ integrated-optic modulators. Ph.D. thesis, University of California San Diego (1990).
4. *ANSOFT HFSS*^{tm}, software produced by Ansoft Corporation, 225 West Station Square Drive, Suite 200, Pittsburgh, PA 15219. www.ansoft.com.
5. M. M. Howerton and W. K. Burns, Broadband traveling wave modulators in LiNbO. Chapter 5 in *RF Photonic Technology in Optical Fiber Links*, ed. W. S. C. Chang, Cambridge University Press (2002).
6. O. G. Ramer, Integrated optic electro-optic modulator electrode analysis. *IEEE J. Quant. Elect.*, **QE-18** (1982) 386.
7. Haeyang Chung, W. S. C. Chang, and E. L. Adler, Modeling and optimization of traveling-wave LiNbO$_3$ interferometric modulators. *IEEE J. Quant. Elect.*, **27** (1991) 608.

5   Planar waveguide devices

Fields of planar guided waves are confined in the depth direction (designated as the x direction in this book) to the vicinity of the high index layer which is the core. The mathematical description of the planar waveguide modes has already been discussed in Sections 1.2.3 and 1.2.4. Since the high index layer is often located near the surface, the guided waves are sometimes called surface waves. As the surface contour of the various layers of the waveguide changes gently, the planar guided-waves will follow the contour. Planar guided waves have three distinct properties.

(1) The evanescent field of the guided-wave modes extends into the air (or cladding) above the surface. Thus they can be excited or coupled out of the core from the air (or cladding layer) adjacent to the surface.
(2) The scattered radiation of the propagating wave is often also visible in the free space above. It can be used to monitor the propagation of the guided wave.
(3) Guided waves are free to propagate in any direction in the transverse plane (designated as the yz plane in this book).

Summation of planar guided waves can form divergent, convergent or diffracted waves in the transverse plane. How to analyze the generalized planar guided waves has already been discussed in Section 1.2.5.

A distinct feature of planar waveguide devices is the utilization of the diffraction, focusing and collimation properties in the transverse plane to achieve focusing, switching, deflection, wavelength filtering or other functions. For example, similarly to wave propagation in the free space, transverse aperture restriction will produce diffraction effects such as radiation lobes. The wider the guided-wave beam, the narrower is the main diffraction lobe. Lenses can be used to focus or collimate guided waves. A surface grating will diffract guided waves into a different direction of propagation. The wavelength selective property of the grating provides the wavelength filtering function. Acousto-optical diffraction in the yz plane can be used for signal processing and beam scanning. How to harness planar guided waves in the transverse plane, and some of the relevant devices, will be the focus of discussion in this chapter.

5.1   Excitation and detection of planar guided waves

5.1.1   End excitation

If a planar waveguide has a cleaved or polished end surface, radiation from a laser or optical fiber can be used to excite planar guided waves by end excitation. Section 1.2.7

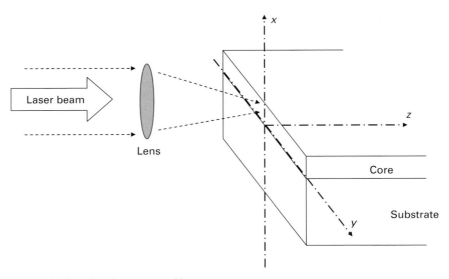

Fig. 5.1.   End excitation of a planar waveguide.

discussed the end excitation of a channel waveguide mode. Efficient excitation of a planar waveguide mode requires that the incident beam be aligned in the direction of propagation with respect to the direction of propagation of the waveguide mode, and that the beam pattern matches the guided-wave mode pattern in the thickness direction. Figure 5.1 illustrates excitation of a waveguide which has a vertical end surface by a laser beam. In order to match the amplitude pattern of the laser beam with that of the TE_0 guided-wave mode in the x direction, a lens is used to obtain the desired focused size of the laser beam.[1] Excitation of TE modes requires the incident electric field to be polarized in the y direction parallel to the core layer, while excitation of the TM modes requires the electric field to be polarized in the x direction, perpendicular to the core layer. The horizontal variation of the incident beam such as the beam size and direction determine the propagation pattern in the transverse plane.[2] For example, the TE_0 or TM_0 guided wave excited by a sharply focused laser or fiber is usually a divergent beam.

Mathematically, for convenience sake, let us assume that the TEM electric field from a laser or optical fiber is incident on the vertical end of the waveguide in the z direction. Its electric field in the xy plane at $z = 0$ is $E_{inc}(x,y)$. At $z \geq 0$, the electric field in the waveguide excited by E_{inc} can be expressed as a summation of all the modes. Let $E_{inc}(x,y) = E_{inc}(x) E_{inc}(y)$, then we can express the summation of modes in the x and y direction separately at $z = 0$. In the x direction, we have

$$E_{inc}(x) = \sum_m A_m \psi_m(x) + \int_\beta b(\beta_x)\psi(\beta_x;x)d\beta_x, \qquad (5.1)$$

where $\psi_m(x)$ are the mth order planar guided-wave modes that are above cut-off, and A_m are their amplitudes. Note that $\psi(\beta_x)$ are the continuous radiation modes (i.e. substrate and air modes) with propagation constant β_x in the x direction, and $b(\beta_x)$ are their amplitudes. Using the orthogonal properties of modes, we have shown in Eq. (1.51) that

$$|A_m| = \frac{\left| \int\limits_{-\infty}^{+\infty} E_{\text{inc}}(x)\psi_m(x)\mathrm{d}x \right|}{\int\limits_{-\infty}^{+\infty} \psi_m(x)\psi_m^*(x)\mathrm{d}x}. \tag{5.2}$$

The power efficiency of exciting the mth mode is $|A_m|^2 \left[\int\limits_{-\infty}^{+\infty} \psi_m \psi_m^* \mathrm{d}x \middle/ \int\limits_{-\infty}^{+\infty} E_{\text{inc}}(x) E_{\text{inc}}^*(x)\mathrm{d}x \right]^2$. Note that $|A_m|$ is large when the overlap integral, $\left| \int\limits_{-\infty}^{+\infty} E_{\text{inc}}(x)\psi_m(x)\mathrm{d}x \right|$, is large. High excitation efficiency implies that the incident radiation is well matched in both amplitude and phase to the mth guided-wave mode.[3] Although Eq. (5.2) is applicable to all the modes, only the excitation of TE$_0$ or TM$_0$ mode is important in many practical applications. Since ψ_0 has a constant phase variation in x, E_{inc} is phase matched only to the $m=0$ mode when it has a uniform phase in the x direction. In addition, the amplitude variation of E_{inc} needs to be matched to that of ψ_0. When E_{inc} is not matched well to any guided-wave mode, it will excite a lot of radiation modes, i.e. $b(\beta_x)$ will have significant amplitudes. In practice, the calculation of $b(\beta_x)$ can be quite complex. Since we are interested usually only in how the guided-wave modes can be excited efficiently, and since the radiation modes are radiated away at $z \gg 0$, the weak radiation field is often not calculated.

In order to determine the diffracted pattern of guided waves in the transverse yz plane, we express $E_{\text{inc}}(y)$ at $z=0$ in terms of a summation of generalized planar mth guided-wave modes. For the mth order guided wave (which has a specific field variation in the x direction and an effective index $n_{\text{eff},m}$), $E_{\text{inc}}(y)$ can be expressed as

$$E_{\text{inc}}(y) = \int\limits_{-n_{\text{eff},m}k}^{+n_{\text{eff},m}k} B_m(\beta_y)\mathrm{e}^{-\mathrm{j}\beta_y y}\mathrm{d}\beta_y , \tag{5.3}$$

where $B_m \mathrm{e}^{-\mathrm{j}\beta_y y}\mathrm{e}^{-\mathrm{j}\beta_z z} = B_m \mathrm{e}^{-\mathrm{j}n_{\text{eff},m}k \sin\theta y}\mathrm{e}^{-\mathrm{j}n_{\text{eff},m} \cos\theta z}$ is a planar guided wave propagating in a direction θ from the z axis. We can find $B_m(\beta_y)$ by multiplying the above equation by $\mathrm{e}^{+\mathrm{j}\beta_y' y}$ and integrating from $-\infty$ to $+\infty$ with respect to y to obtain

$$\int\limits_{-\infty}^{+\infty} E_{\text{inc}}(y)\,\mathrm{e}^{+\mathrm{j}\beta_y' y}\mathrm{d}y = \int\limits_{-\infty}^{+\infty} \left(\int\limits_{-n_{\text{eff},m}k}^{+n_{\text{eff},m}k} B_m(\beta_y)\mathrm{e}^{-\mathrm{j}\beta_y y}\mathrm{d}\beta_y \right) \mathrm{e}^{+\mathrm{j}\beta_y' y}\mathrm{d}y$$

$$= \int\limits_{-n_{\text{eff},m}k}^{+n_{\text{eff},m}k} B_m(\beta_y) \left(\int\limits_{-\infty}^{+\infty} \mathrm{e}^{-\mathrm{j}\beta_y y}\mathrm{e}^{+\mathrm{j}\beta_y' y}\mathrm{d}y \right) \mathrm{d}\beta_y. \tag{5.4}$$

From Fourier Transform theory, we know

$$\frac{1}{2\pi} \int_{-\infty}^{+\infty} e^{-j\beta_y y} e^{+j\beta'_y y} dy = \delta\left(\beta_y - \beta'_y\right). \tag{5.5}$$

Therefore,

$$B_m\left(\beta_y\right) = \frac{1}{2\pi} \int_{-\infty}^{+\infty} E_{\text{inc}}(y) e^{+j\beta_y y} dy. \tag{5.6}$$

The electric field of the mth guided-wave mode in x, y and z, excited by E_{inc} is

$$E_m(x, y, z) = A_m \left[\int_{-n_{\text{eff},m}k}^{+n_{\text{eff},m}k} B_m\left(\beta_y\right) e^{-j\beta_y y} e^{-j\beta_z z} d\beta_y \right] \psi_m(x) = E_m(y, z) E_m(x). \tag{5.7}$$

We can also find $E_m(y,z)$ by solving the two-dimensional Helmholtz equation in Eq. (1.39) and (1.41) where the boundary condition is given by the incident field $E_{\text{inc}}(y)$ at $z = 0$ [1, 2].

In order to demonstrate the analysis, let us consider a simple example where $E_{\text{inc}}(y)$ has uniform amplitude A for $|y| \leq l_y$ and zero for $|y| > l_y$. Then

$$B_m = \frac{1}{\pi} A l_y \frac{\sin\left(\beta_y l_y\right)}{\beta_y l_y}, \tag{5.8}$$

where B_m is the amplitude of the planar guided wave propagating in the θ direction and excited by E_{inc}. The value of B_m will be large primarily at $n_{m,\text{eff}} k \sin\theta l_y \ll \pi$. B_m is 0 at $n_{m,\text{eff}} k \sin\theta l_y = \pi$. Usually $l_y \gg \lambda$ where $\lambda = 2\pi/k$. It means that only planar guided waves with a small angle θ with respect to the z axis will have a large amplitude which is proportional to $\sin\left(\frac{2\pi n_{\text{eff},m} l_y}{\lambda} \sin\theta\right) / \frac{2\pi n_{\text{eff},m} l_y}{\lambda} \sin\theta$. Those planar guided waves constitute the main lobe excited by E_{inc}. The center of the main lobe is directed toward $\theta = 0$. When θ increases, the amplitude of the planar guided wave drops to 0 at $\theta_o \approx \sin\theta_o = \lambda/2n_{\text{eff},m} l_y$. The θ_o is controlled by l_y. The smaller the l_y, the larger is the θ_o. Such a diffraction pattern is similar to the diffraction pattern of a plane wave by a rectangular aperture in free space.

Note that if E_{inc} is incident on the waveguide at an angle δ with respect to the z axis in the yz plane, it would have a functional variation $\exp(-jk \sin\delta y)$ at $z = 0$. In that case, the result obtained from Eq. (5.6) will be

$$B_m = \frac{1}{\pi} A l_y \frac{\sin\left((\beta_y - k\sin\delta)l_y\right)}{(\beta_y - k\sin\delta)l_y}. \tag{5.9}$$

It means that the center of the main lobe will be directed at $\theta = \sin^{-1}\left(\frac{1}{n_{\text{eff},m}} \sin\delta\right)$. This is just Snell's law of refraction.

The major advantage of end excitation is its simplicity. However, the waveguide must have a smooth end surface obtained usually by either cleaving or polishing. For a given incident radiation all modes that have finite overlap integral with the incident field will *be excited. The only way to control the excitation of the modes is by the polarization and by the field pattern of the incident radiation that controls the overlap integral. This property has important practical implications. For example: (1) if a laser beam focused through a spherical lens is used to excite a TE_0 or TM_0 mode, the spot size focused on the end of the waveguide should have a flat wave-front and a spot size matching the mode size in the x direction. It implies that the spot size will also be small in the y direction, creating a divergent beam in the yz plane. (2) A TE_1 mode will not be excited by an E_{inc} field symmetric with respect to the center of the waveguide.*

5.1.2 Excitation by prism coupler

5.1.2.1 Interaction of waves in a prism and a planar waveguide

Let there be a prism surface parallel to the yz plane and a planar waveguide with top surface parallel to the prism surface. The planar waveguide is placed below the prism. They are separated by an air gap g in the x direction. Figure 5.2 illustrates this config-uration. When an incident plane wave propagates at an angle $-\theta_p$ with respect to the z axis in a prism that has a refractive index n_p, the propagation wave constant in the $+z$ direction is $n_p k \cos \theta_p$. Let n_p be larger than the effective index of the planar guided-wave mode. Then $n_p \cos \theta_p = n_{eff,m}$ can be achieved at a specific angle θ_{pm}. Since $n_p > n_{eff,m} \gg 1$, the incident plane wave at $-\theta_{pm}$ is totally internally reflected at the prism surface for large air gaps. The reflected wave is a plane wave at angle $+\theta_{pm}$. Both the incident and the reflected plane wave in the prism will propagate toward the $+z$ direction with an $\exp(-jn_p k \cos \theta_p)$ variation. They have an evanescent field in the $-x$ direction, with an $\exp\left[\left(\sqrt{n_p \cos \theta_p^2 - 1}\right)k(x - g)\right]$ variation in the air gap.

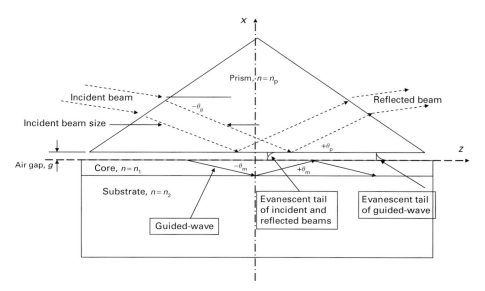

Fig. 5.2.   Coupling of waves in a prism and a waveguide.

The mth planar guided wave propagating in the z direction will have a z variation of $\exp(n_{m,\mathrm{eff}}kz)$ and an $\exp\left[-\left(\sqrt{n_{\mathrm{eff},m}^2 - 1}\right)kx\right]$ variation in the air gap. Let the guided wave have a significant evanescent field in the air gap. At $\theta_\mathrm{p} = \theta_\mathrm{pm}$, the evanescent field of the plane waves propagates in synchronism with the evanescent tail of the mth order guided wave mode. When the distance of separation g, i.e. the air gap between the prism surface and the waveguide surface, is less than or comparable to the length of the evanescent tail, traveling wave interaction between the plane waves and the forward propagating planar guided-wave mode takes place. The incident plane wave excites the guided-wave mode while the guided-wave mode adds to the reflected plane wave. For an incident plane wave beam that has a finite size in the z direction, the amplitude of the guided wave as a function of z will first grow as the incident wave transfers its power into the guided wave. The amplitude of the guided-wave mode is too small in the beginning to transfer much power back to the reflected wave. As the guided wave grows stronger, more power is transmitted to the reflected wave. Finally, after the end of the incident beam, all power in the guided wave will eventually be transferred to the reflected beam. In summary, the amplitude of the guided wave will first grow and then decay. The total reflected beam will continue after the end of the incident beam until the guided-wave power is exhausted.

The amplitude of the reflected beams at θ_pm in the prism and the mth guided wave mode as a function of z can be analyzed by considering the incident and reflected beams to be composed of a series of parallel optical rays at different z positions. Each optical ray is still much wider than the optical wavelength so that it can be analyzed approximately as a plane wave. The guided-wave modes are also represented as a series of rays of plane waves at different z positions, propagating at their specific angles of propagation $\pm\theta_m$ in the core and reflected at the boundaries of the core as discussed in Sections 1.2.3.4 and 1.2.4.4. Each incident ray at $-\theta_\mathrm{pm}$ in the prism and at a given position z is partially reflected at the prism surface and partially transmitted to the ray of the plane wave of the mth waveguide mode at $-\theta_m$ in the core. The transmitted energy adds on to the amplitude of the existing plane wave for the mth order waveguide mode in the core.[4] The plane wave of the mth waveguide mode at $+\theta_m$ is partially reflected at the top boundary back to the core and partially transmitted to the reflected ray in the prism through the air gap. Thus the total amplitude of the reflected ray in the prism at the position z is the sum of the reflection at the prism boundary and the transmission from the guided wave. In summary, the guided wave receives energy from the incident ray, it also leaks energy to the reflected ray in the prism.[5]

A mathematical analysis of the cumulative effect of all the optical rays yields the amplitude of the reflected beam and the guided-wave as a function of z [3, 4]. It is straightforward, but long, and so it is not repeated here. It has been shown from such analysis that: (a) the reflected optical rays in the prism obtained from the prism boundary reflection and from the guided-wave mode are 180° out of phase. So the reflection first gets weaker after the beginning of the incident beam. The total reflected ray in the prism reaches zero at some distance after the leading edge of the incident beam. After that the reflected wave transmitted from the waveguide mode dominates. The total amplitude of the reflected ray keeps increasing until the end of the incident optical beam. After the incident beam has ended, the total amplitude of the reflected rays then decays to zero within a short distance as the power in the guided wave decays to zero. In short, the reflected beam looks as if it has been shifted in position from the incident beam. This is called the Goos–Haenchen shift of the prism coupler. (b) The amplitude of the guided wave increases until the end of the incident beam, then it decays. (c) For a given air gap and length of the evanescent field tail, there is an optimum width of the incident optical beam at which the efficiency of power transfer from the incident beam to the mth order guided wave is a maximum at the end of the incident beam. At this optimum beam width, the maximum excitation efficiency is 81%. If the leakage of optical power from the guided wave to the reflected beam is stopped at this z position, then the power can be retained in the guided-wave mode. (d) As the θ_p of the incident beam increases from zero, it will excite first the zeroth order mode, then the first order mode, etc. The polarization of the incident beam determines whether the TE or the TM modes will be excited. (e) If there is power in all the propagating modes in the waveguide and a prism is placed on top of the waveguide with sufficiently small air gap, then power in different modes will be coupled out as optical beams propagating at different angles θ_{pm}. If a screen is placed at the output of the prism, these outputs at each specific θ_{pm} will appear as illuminated lines, called the m-lines. They will appear as m-lines instead of m-dots because the guided-wave beam spreads in the horizontal direction.

5.1.2.2   The prism coupler

Theoretically, for efficient excitation of the guided wave, we need a prism that has a refractive index larger than the effective index of the mode. We also need an air gap comparable to (or less than) the length of the evanescent tail.[6] We should adjust the size of the incident beam for a given air gap to achieve the maximum efficiency of 81% at the end of the incident beam. Simultaneously, we should terminate the coupling between the prism and the waveguide at the end of the incident beam. Experimentally, it is difficult to control the width of the incident beam in the z direction in coordination with adjustment of air gap separation to obtain maximum coupling efficiency. It is even difficult to obtain a pre-specified air gap separation. In practice, there are often settled dust particles or defects on either the prism or the waveguide surface. The size of these particles or defects is comparable to or larger than the desired range of air gap separation. There is no effective way to monitor the air gap separation.

For these reasons, a right angled prism such as that illustrated in Fig. 5.3(a) with index $n_p > n_{m,\text{eff}}$ is used experimentally to excite the guided wave. Let there be an incident beam of finite size. The size of the incident beam is controlled by a lens. The incident beam is first adjusted in angle of incidence with respect to the prism to obtain the desired θ_{pm}. Its effective excitation length in the z direction is also cut short by the vertical edge of the prism. In order to adjust the air gap, a localized pressure is exerted by a wedge or a ball point on the substrate of the waveguide toward the prism, near the vertex of the prism, by a mechanical jig. The air gap is determined by the size of the random dust particles or surface irregularities and the pressure. The larger the pressure, the smaller is the air gap. The prism and the waveguide now form an assembly with the air gap set by the pressure. Then the waveguide together with the prism assembly is slid in the z direction toward the

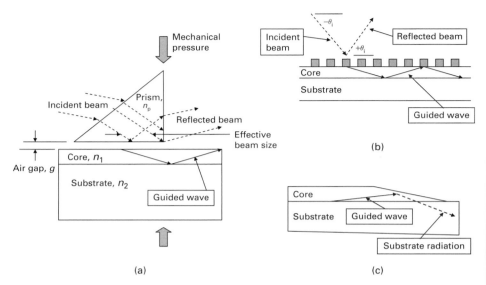

Fig. 5.3.   Couplers for planar waveguides. (a) A prism coupler. (b) A grating coupler. (c) A tapered thin film coupler.

incident beam so that the vertex of the prism cuts into the incident beam. The effective size of the incident beam exciting the guided wave is now between the leading edge of the incident beam and the edge of the right angle prism. The intensity of the guided wave is monitored at a z position after the vertex of the prism.[7] At an optimum combination of the effective beam size and the mechanical pressure exerted on the prism assembly, the monitored intensity of the guided wave will reach a maximum. This is the maximum power that can be transferred from the effective portion of the incident beam to the mth order guided wave for this air gap. Note that when the effective beam size is significantly smaller than the size of the incident beam, the power in the part of the incident beam cut off by the vertical edge of the prism is lost. Therefore, even when the excitation of the guided wave for the effective portion of the incident beam has reached 81%, the efficiency of excitation for the entire incident beam may be much lower. In order to see whether the intensity of the guided wave can be further increased, the localized pressure exerted on the substrate is adjusted to obtain a new air gap, and the incident beam is slid against the vertical edge of the prism until a new maximum intensity of the monitored guided wave is obtained. The intensity of the guided wave is monitored and compared with the intensity obtained in the previous round. The pressure on the waveguide substrate can be adjusted again until the desired excitation efficiency is obtained. Obviously, the size of the incident beam could also be adjusted.

A major advantage of the prism coupler is that each mode is excited by the incident radiation only at the specific angle. The excitation can be initiated at any position by the placement of the prism. The direction of the excited guided wave is oriented along the direction of the incident beam. Since the excitation efficiency could be high for relatively large incident beam size, e.g. a millimeter or more, the beam divergence of such a wide guided-wave beam in the yz plane will be small.

5.1.3   The grating coupler

An incident plane wave propagating at angle $-\theta_i$ with respect to the z axis will have a z variation $\exp(-jk\cos\theta_i)$. When there is a grating with a periodicity T that satisfies the Bragg condition

$$\frac{2\pi}{T} = n_{\text{eff},m} \pm k\cos\theta_i,$$

phase matched interaction between the incident plane wave and the mth guided wave will occur. The grating will diffract energy from the incident wave into the guided wave. Vice versa, the grating will also diffract energy from the guided wave into an output radiation beam.[8] This is illustrated in Fig. 5.3(b). It is similar to a prism coupler and is known as a grating input coupler. Its analysis is very similar to that of the prism coupler [5, 6].

However, the grating may have more than one order of diffraction. The diffraction condition stated in the equation above is for the first order. The efficiency of a grating coupler is affected by its higher orders of diffraction. A more detailed analysis of diffraction by a grating will be presented in Section 5.2.1.1(C). *In summary, the grating coupler functions like a prism coupler without the cumbersome prism coupler assembly. However, the grating needs to be fabricated on the waveguide. The coupling efficiency is reduced by higher order diffractions.*

5.1.4   The tapered waveguide coupler

Let there be a propagating mth order mode in the waveguide. When the thickness of the core is reduced, it will eventually reach the cut-off thickness of the mode at a certain point.[9] When the propagating mode reaches that point, all the power in the propagating mode will be transferred to the substrate modes. The substrate modes will then form a radiation beam in the substrate as illustrated in Fig. 5.3(c). Therefore, the power in the mth order mode is coupled out of the waveguide from the substrate, known as the tapered thin film coupler. In principle, by reciprocity, a thin film coupler could also be used as an input coupler, provided that the incident beam has the specific beam pattern that excites the guided-wave mode effectively. *Since the shaping of the input beam in the substrate into a specific pattern is difficult, the tapered thin film coupler has been used only as an output coupler.*

5.1.5   Detection and monitoring of guided waves

The most common method for detection and monitoring of planar guided waves is to use a right angled output prism with refractive index n_p larger than the $n_{m,\text{eff}}$ of the guided-wave mode as illustrated in Fig. 5.4. The prism is mechanically pressed against the waveguide to obtain an air gap comparable to the evanescent tail length, similarly to the assembly of the input prism coupler. For any reasonable air gap, all of the energy (minus scattering loss) in the mth guided-wave mode is coupled out as an output beam at the angle θ_{pm} in the prism (i.e. the m-line). The larger the air gap, the wider is the output

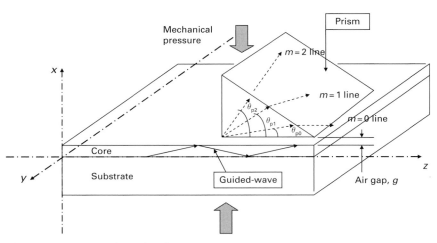

Fig. 5.4.   The prism output coupler and *m*-lines.

beam, and the narrower is the angular spread of the output beam. The number of *m*-lines represents the number of modes that have been excited in the waveguide. The *m*-lines are illustrated in Fig. 5.4. *The use of a prism for output coupling has an important advantage: each mode will appear only as a discrete m-line at θ_{pm}. When a receiver detects the output of a specific m-line, it measures only the energy in the mth order guided-wave mode. The output coupling efficiency is close to 100% for each mode. It is easy to use experimentally.* Vice versa, stray light propagating as substrate and air modes in the waveguide will be coupled out at other angles. Therefore the receiver measures accurately the energy in the *m*th guided-wave mode, no matter how many modes are propagating.

The output prism could also be used to measure the propagation loss of the *m*th order mode in that waveguide. When the guided-wave mode propagates with attenuation coefficient α, its intensity will vary as $\exp(-\alpha z)$. When the intensity of the *m*-line is plotted as a function of the z position of the output prism on a logarithmic scale, it will be a straight line, and the slope of the straight line is the measured attenuation coefficient α of the *m*th mode.[10]

Some waveguides have a smooth vertical end fabricated by cleaving or polishing. In that case, the radiation pattern of the propagating modes at the output end can be directly observed, imaged, or detected. However, if there are several guided-wave modes, they will appear simultaneously. Unless there is only one mode, it may be very difficult to determine what are the excited modes and their relative strength. Fortunately, we know from theoretical analysis the radiation pattern of the modes. When there are just very few modes, we may be able to identify the modes by recognition. Moreover, stray radiation (i.e. substrate and air modes) will usually appear on the end surface at locations far away from the expected pattern of the guided-wave mode.

Let there be only one propagating mode in the waveguide, and let there be different lengths L of identical waveguides. The ratio of the output to input intensity of the mode will again vary as $\exp(-\alpha L)$. Let us excite the guided-wave mode and measure the output intensity separately for each waveguide that has different L. When the excitation efficiency can be the same for waveguides of different L, then the α of this mode is the slope of the logarithmic plot of the ratio of detected output intensity/laser input versus L. This is called the cut back method for the measurement of propagation loss. However, there will be variation of the excitation efficiency. Therefore, there will be scattering of the measured output intensity. The measured slope gives only an averaged α, and it is reliable only when α is larger than the scattering of the data points.

For waveguides that have considerable scattering loss, the scattered radiation may be easily detected in the free space above the waveguide. The α can also be obtained by plotting the detected scattered radiation as a function of the z position. This method is accurate only if the same fraction of the scattered radiation can be detected as the z position is varied, for example by a microscope focused on a well-collimated planar guided-wave mode. However, it is the simplest method for monitoring the power in the waveguide and for measuring the α.

5.2 Diffraction, focusing, and collimation in planar waveguides

Ability to focus, refract, diffract, filter, and collimate guided waves in the transverse plane is very important for many applications in planar waveguides.

5.2.1 The diffraction grating

Gratings are fabricated on planar waveguides either by etching the grating pattern on to the cladding layer or on to the core. A grating can also be obtained by depositing a material that has the grating pattern on the waveguide. When an electrode with a grating pattern is fabricated on an electro-optic active waveguide such as $LiNbO_3$ or a polymer and when a voltage is applied to the electrode, the electro-optic effect of the applied electric field creates a periodic change of refractive index in the form of a grating. In Section 3.4, a time varying periodic change of refractive index produced by an acoustic surface wave was discussed.

An ideal static etched or deposited grating would have a periodic rectangular spatial profile for the grooves which have permittivity ε', periodicity T, groove width δ, thickness d, and groove length W, as illustrated in Fig. 5.5. Note that $\Delta\varepsilon$ is the spatial variation of the permittivity between the total waveguide structure with and without the grating. In other words, when a grating is fabricated, the original waveguide is perturbed by the $\Delta\varepsilon$. Mathematically, the $\Delta\varepsilon$ of the grating is a periodic summation of all the grooves,

$$\Delta\varepsilon(x,y,z) = \left[\sum_m (\varepsilon' - \varepsilon_0)\,\text{rect}\left(\frac{mT-z}{\delta/2}\right)\right]\text{rect}\left(\frac{x-H}{d/2}\right)\text{rect}\left(\frac{W/2-y}{W/2}\right),$$

where

$$\text{rect}(\tau) = 1 \quad \text{for} \quad |\tau| \leq 1 \quad \text{and} \quad \text{rect}(\tau) = 0 \quad \text{for} \quad |\tau| > 1.$$

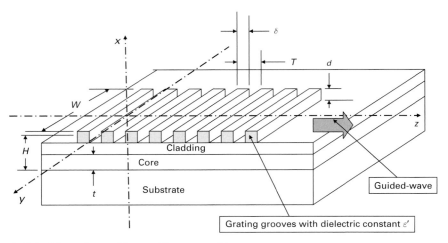

Fig. 5.5.   A dielectric grating co-linear with a waveguide.

Here ε_0 is the free space permittivity. In reality, the spatial profile of the groove may also be trapezoidal or triangular. For gratings with rectangular grooves, $\Delta\varepsilon(x,y,z) = \Delta\varepsilon(z)\Delta\varepsilon(x,y)$. It is well known that any periodic function of z can always be represented by its Fourier series. Since the grating in Fig. 5.5 is an even function of z, we have

$$\Delta\varepsilon(z) = \frac{1}{T}\int_0^T \Delta\varepsilon\,dz + 2\left(\int_0^T \Delta\varepsilon\cos Kz\,dz\right)\cos Kz + 2\left(\int_0^T \Delta\varepsilon\cos 2Kz\,dz\right)\cos 2Kz$$

$$+ 2\left(\int_0^T \Delta\varepsilon\cos 3Kz\,dz\right)\cos 3Kz + \text{higher orders.} \tag{5.10}$$

The fundamental term of the Fourier series which represents $\Delta\varepsilon(z)\Delta\varepsilon(x,y)$ is

$$\Delta\varepsilon_0\,\cos(Kz)\,\text{rect}\left(\frac{x-H}{d/2}\right)\text{rect}\left(\frac{W/2-y}{W/2}\right). \tag{5.11}$$

This sinusoidal grating has a periodicity $T = 2\pi/K$ in the z direction, a width W in the y direction, a thickness d in the x direction, and a maximum change of dielectric constant $\Delta\varepsilon_0$, $\Delta\varepsilon_0 = 2\int_0^T \Delta\varepsilon(z)\cos Kz\,dz$. The $\Delta\varepsilon$ perturbation layer is centered at $x = H$, where $H \geq t+(d/2)$. There are also higher order terms with declining magnitudes at periodicity $T/2$, $T/3$, $T/4$, $T/5$, etc. The relative magnitude of the various Fourier terms will be determined by the shape of the grooves.

5.2.1.1   Co-linear diffraction
(A) The reflection wavelength filter
Let there be a planar mth order guided wave propagating along the z direction of the grating as shown in Fig. 5.5. The grating is placed in the evanescent field of the mth order guided-wave mode. For example, a grating may be deposited on the cladding layer or on

top of the waveguide. Discussion in Section 2.2.3 has shown that a forward propagating wave in the z direction will be reflected by the grating $\Delta\varepsilon$ in Eq. (5.11) into the backward direction when the phase matching condition

$$K = 2n_{\text{eff},m}\beta_0 = 2n_{\text{eff},m}\left(\frac{2\pi}{\lambda_g}\right), \tag{5.12a}$$

is satisfied.[11] Equation (5.12a) is known as the Bragg condition of reflection which determines the center wavelength at which the maximum reflection takes place. There will also be reflection at wavelengths slightly different than the center wavelength. The phase matched interaction has a pass band within the wavelength range $\Delta\lambda_g = \pm 4\pi C_g n_{\text{eff},m}/K^2$ where C_g is the overlap integral between the $\Delta\varepsilon_0$ of the grating and the guided-wave mode given in Eq. (2.15). The magnitudes of the transmitted and reflected waves are controlled by the length of the grating and the magnitude of C_g.

There are also higher order terms in the Fourier series in Eq. (5.10). The Bragg condition for the nth higher order terms is

$$nK = \frac{2n\pi}{T} = 2n_{\text{eff},m}\beta_0 = 2n_{\text{eff},m}\left(\frac{2\pi}{\lambda_{gn}}\right). \tag{5.12b}$$

When this condition is satisfied, reflection of guided waves at wavelengths around λ_{gn} also takes place. Note that for a given wavelength, Bragg reflection at higher orders takes place at much larger periodicity T. A grating with large T is much easier to fabricate, but the coupling coefficient C_g is smaller.

In general, the co-linear interaction is used primarily as a wavelength filter. A reflection grating filter will usually be designed so that only signals within a desired band of wavelength will be reflected. An important observation for planar guided waves is that Bragg reflection takes place even when the guided wave is incident on to the grating at a moderate angle ξ with respect to the z axis. In that case the Bragg condition is

$$nK = 2n_{\text{eff},m}\beta_0 \cos\xi = 2n_{\text{eff},m}\left(\frac{2\pi}{\lambda_g}\right)\cos\xi. \tag{5.12c}$$

The reflected beam will be in the $-\xi$ direction. It implies that the incident beam will be reflected into a new direction like a mirror.

(B) Reflection and transmission between different modes

Let the waveguide have multimodes. A coupled mode analysis similar to those shown in Section 2.2.3 will show that an incident mth order mode propagating in the $+z$ direction will transfer part or all of its power to the nth order mode when

$$K = \frac{2\pi}{T} = \left|n_{\text{eff},m} \pm n_{\text{eff},n}\right|k = \left|n_{\text{eff},m} \pm n_{\text{eff},n}\right|\frac{2\pi}{\lambda_g}, \tag{5.13}$$

where the $-$ sign applies to a forward propagating nth order mode, and the $+$ sign applies to the reverse propagating nth mode. However, there is a significant difference in the power transfer characteristics dependent on the sign.

In the case of a forward mth order mode interacting with the reverse nth order mode, the coupled mode equation equivalent to Eq. (2.15) will be

$$\frac{da_f}{dz} = -ja_b \frac{C_g}{2} a_b e^{j(n_{\text{eff},m}k + n_{\text{eff},n}k - K)},$$

$$\frac{da_b}{dz} = +ja_f \frac{C_g}{2} a_f e^{j(n_{\text{eff},m}k + n_{\text{eff},n}k - K)},$$

$$C_g = \frac{\omega}{4} \int_{H-\frac{d}{2}}^{H+\frac{d}{2}} \Delta\varepsilon_0 \left| \underline{e_m} \bullet \underline{e_n^*} \right| dx, \tag{5.14}$$

where W is assumed to cover the entire width of the guided-wave beam. Solutions of the amplitude of the forward and backward waves in Eq. (5.14) are in hyperbolic sine and cosine functions as expressed in Eq. (2.17). It means that the longer the grating, the stronger is the reflection. The larger the $\Delta\varepsilon_0$ and d, the wider is the wavelength band for effective reflection. The reflection characteristics will be similar to those of the reflection grating discussed in Section 2.2.3. The main objective of the grating is to reflect the power in the mth order mode into an nth order mode within a certain desired wavelength bandwidth, i.e. a reflection filter.

In the case of a forward mth order mode interacting with a forward nth order mode, the coupled mode equation will have a $-$ sign on the right hand side of the equations for both da_f/dz and da_b/dz. It leads to solutions of a_f and a_b in cosine and sine functions, similar to those shown in Eq. (3.47) when $\Omega = 0$. It means that the power transfer between the two modes will be periodic in z. The larger the $\Delta\varepsilon_0$ and d, the shorter is the period. When the phase match condition in Eq. (5.13) is satisfied, 100% transfer of power can be obtained whenever $C_g W = \pi/2$, $3\pi/2$, $5\pi/2$, etc. The usual objective of the grating is to transfer a specific amount of the power from the mth order mode into the nth order mode.

(C) Coupling to air and substrate radiation – the grating coupler

When the K of the grating $\Delta\varepsilon$ provides the phase match condition between the forward propagating mth guided-wave mode and an air or substrate mode with propagation constant β in the z direction, through the nth order Fourier series, we have, analogously to Eq. (5.12), the Bragg condition of diffraction

$$nK = \left| n_{\text{eff},m}k \pm \beta \right| = \left| n_{\text{eff},m} \frac{2\pi}{\lambda_g} \pm \beta \right|, \tag{5.15}$$

where the $+$ sign applies again to substrate (or air) modes in the $-z$ direction, and the $-$ sign applies to the substrate (or air) modes in the $+z$ direction. Note that $\beta = n\cos\theta$ where n is the index of the substrate (or air) and θ is the propagation angle of the radiation from the z axis in the substrate (or air).[12] However, the coupled mode equation does not apply. The analysis of the power transfer will now be similar to that of a prism coupler discussed in Section 5.1.2.2. When there is any incident substrate (or air) mode at the phase matching angle, the guided-wave mode receives power from the incident radiation mode and leaks energy to all the diffracted substrate and air modes that satisfy Eq. (5.15).

There is no coupling of energy back to the incident radiation mode or the guided-wave mode from the diffracted outgoing waves. The grating has been used as input coupler and output coupler like the prism coupler.

A grating coupler is more difficult to fabricate, but it is more convenient to use without the prism assembly. In a grating output coupler, when there is no incident substrate mode and when there is power in the mth guided-wave mode, the grating enables transfer of power into all the substrate (or air) radiation modes at all the phase matched angles. Note that, in an input coupler, when there is only one substrate mode with β that satisfies Eq. (5.12), there will be no loss of power through other orders of diffraction. Then the maximum excitation efficiency of a grating input coupler is also 81% [6]. If there are substrate or air modes that are phase matched to the guided-wave mode through other orders of diffraction, the excitation efficiency will be much lower.

A practical periodic dielectric grating usually has higher order terms in its Fourier series expansion. The period of those higher order terms will be $1/n$ of the T of the fundamental term shown in Eq. (5.11). There are occasions in which the required T for the fundamental term is too small to fabricate. Then one may fabricate a grating with a period nT and utilize the nth order term for phase matching. An example of this is the use of the 3rd order term for reflection in semiconductor lasers. In that case, since the material refractive index for semiconductor lasers is more than 3, gratings with periodicity that satisfy Eq. (5.12) in the first order will be very difficult to fabricate. However, there will be transfer of power to all those modes that satisfy the phase matching condition through different order terms.[13] The coupling of those modes to the guided-wave mode is a power loss. For this reason, grating input couplers often only have efficiencies in the 30% range. Only gratings that provide phase match just between a forward planar guided wave and a backward substrate mode may have high input efficiency. In such a grating, there is no phase matched transfer of power to any other mode.

5.2.1.2 Deflection by grating

Let there be a grating that has its grooves oriented along the z direction and periodic variation of ε in the y direction with periodicity $T = 2\pi/K$, thickness d, width δ, centered at $x=H$ and length W, located from $z=0$ to $z=W$, as shown in Fig. 5.6. It is similar to the grating shown in Fig. 5.5, except that its orientation is rotated to the y direction. The fundamental term of the Fourier series expansion of the dielectric variation of the grating can again be written as

$$\Delta\varepsilon_0 \cos(Ky)\text{rect}\left[\frac{2(x-H)}{d}\right]\text{rect}\left(\frac{W/2-z}{W/2}\right). \tag{5.16}$$

Often, $\boldsymbol{K} = K\,\boldsymbol{i_y}$ is used to designate the periodic variation of its fundamental order. This fundamental term is also similar to the acousto-optical grating in Eq. (3.28) with $\Omega = 0$. Let there also be a planar mth guided-wave propagating at an angle $-\theta_i$ from the z axis incident on the grating. From discussions in Section 3.5.3 and Eq. (3.45), we know that there will be a deflected planar guided-wave propagating at an angle $+\theta_d$ from the z axis when $\underline{\boldsymbol{K}} = \pm\left(\underline{\boldsymbol{\beta}}_i - \underline{\boldsymbol{\beta}}_d\right)$, as illustrated in the inset of Fig. 5.6. In other words,[14]

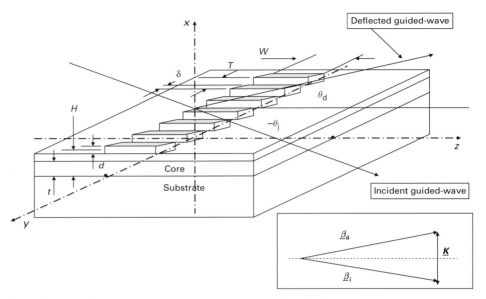

Fig. 5.6.　　A deflection grating. The inset shows the matching of $\underline{K}$, $\underline{\beta}_i$ and $\underline{\beta}_d$.

$$(\sin \theta_i + \sin \theta_d) n_{\text{eff},m} \frac{2\pi}{\lambda_g} = K. \tag{5.17}$$

For a guided wave incident on the grating at $z = 0$, the amplitudes a_i and a_d of the incident and deflected wave at $z \geq W$ are given in Eq. (3.47) to be

$$a_i(W) = A \cos CW, \qquad a_d(W) = A \sin CW, \tag{5.18}$$

where A is the amplitude of the incident wave for $z < 0$, and C is the coupling coefficient

$$C = \frac{\omega}{4} \int_{H-\frac{d}{2}}^{H+\frac{d}{2}} \Delta\varepsilon_0 |\underline{e_m}|^2 \mathrm{d}x. \tag{5.19}$$

Therefore, such a grating is a deflector. Its deflection efficiency is determined by CW. Note that the periodicity T of the deflection grating is much bigger than the period of gratings used for co-linear diffractions, because θ_i and θ_d are usually small. At such large T, electrodes can be fabricated on electro-optic material such as LiNbO$_3$ to create the periodic $\Delta\varepsilon$ pattern. The most common method to obtain such an electro-optic $\Delta\varepsilon$ pattern is to use an interdigital electrode on an electro-optic material such as LiNbO$_3$.

5.2.2　Refraction, collimation, and focusing of planar waveguide modes

5.2.2.1　Refraction of a planar waveguide mode

When the refractive index (or thickness) of either the cladding or the core layers of a waveguide changes from one region to another, the effective indices of the planar guided-wave modes under different regions are different. Let there be a guided wave in the mth mode in region 1 obliquely incident on a straight boundary separating the two regions. The angle of the incident guided wave with respect to the boundary is θ_i. There will also be a transmitted beam across the boundary in region 2 and a reflected beam in region 1. The continuity condition of the tangential electric field must be satisfied along the boundary in both the vertical and the lateral directions. The field of the mth order mode on both sides of the boundary matches very well in the vertical direction. The small difference between the incident and refracted guided waves is made up with radiation fields. In order to match the boundary condition in the lateral direction, the direction of propagation of the mth mode in the adjacent region 2 with respect to the boundary, i.e. θ', must satisfy $n_{\text{eff},m} \cos \theta_i = n'_{\text{eff},m} \cos \theta'$.[15] In other words, Snell's law of refraction is directly applicable using the effective index. For a moderate and smooth discontinuity at the boundary, a negligible amount of power will be diffracted into other guided, substrate or air modes.

A prism for planar guided waves can be made by simply depositing an extra layer of material in the shape of a prism cross-section (i.e. a triangle) on the original waveguide. Note, however, the difference between $n_{\text{eff},m}$ and $n'_{\text{eff},m}$ will be very small. In other words, such a prism is a weak prism.

5.2.2.2   Focusing and collimation of planar waveguide modes

Equation (1.42) shows that any guided wave that has an x variation of the mth order mode and a variation $\exp\left(-jn_{\text{eff},m}k\sqrt{y^2+z^2}\right)$ in the yz plane is an outgoing cylindrical wave radiating from $z = y = 0$. Vice versa, an $\exp\left(+jn_{\text{eff},m}k\sqrt{y^2+z^2}\right)$ variation will represent an incoming cylindrical wave focused on $z = y = 0$. When z is large,

$$\sqrt{z^2+y^2} \cong z + \frac{y^2}{2z}. \tag{5.20}$$

Thus a guided wave at $-z$ that has a phase variation $\exp\left[+jn_{\text{eff},m}k\left(z+\frac{y^2}{2z}\right)\right]$ will be focused at $y = z = 0$. Furthermore, a planar mth order guided-wave mode propagating in the forward z direction will normally have an $\exp\left(-jn_{\text{eff},m}kz\right)$ variation. If its phase front at $z = -f$ can be modified by the factor, $\exp\left(+jn_{\text{eff},m}k\frac{y^2}{2f}\right)$, then it will be converted into a cylindrical wave focused at $y = z = 0$. This is what a waveguide lens placed at $-z$ with a focal length f should do. In other words, an ideal lens with focal length f would transform any input guided wave by multiplying its amplitude variation with a phase factor, $\exp\left(+jn_{\text{eff},m}k\frac{y^2}{2f}\right)$. The representation of a lens by a quadratic phase transformation is commonly used in three-dimensional optical analysis (see Section 1.4.3 of [1]).

Note that when the phase factor $\exp\left(+jn_{\text{eff},m}k\frac{y^2}{2f}\right)$ is applied to an outgoing cylindrical wave at $z = f$, the resultant amplitude variation is $\exp\left(-jn_{\text{eff},m}kz\right)$ which is a planar guided wave in the $+z$ direction. In other words, an outgoing cylindrical guided wave is also collimated by a lens. Needless to say for any lens or guided-wave beam of finite size, there will be a diffraction effect of the limited aperture such as those discussed in Section 5.1.1.

There are several ways to obtain a guided-wave lens, including the Luneberg lens, the geodesic lens and the diffraction lens.

(A) The Luneberg lens

A generalized Luneberg lens in three dimensions is a variable index, circular symmetric refracting structure which re-images two objects to each other. Luneberg and other researchers have analytically determined the refractive index distribution that will give a diffraction limited performance. Using the dispersion relation of the waveguide (i.e. n_{eff} vs. thickness), the analysis has been extended to the required variation of the thickness profile of the waveguide that will yield a waveguide lens [8]. A Luneberg lens has been fabricated by depositing lens material on a planar waveguide through a shaped mask. However, it is difficult to achieve the prescribed effective index distribution. Consequently it has not been used in practice.

(B) The geodesic lens

When a planar waveguide is fabricated on a substrate with a contoured surface, propagation of a guided-wave beam will follow the contour. Let there be a contoured depressed area. Guided-wave beams propagating through the depressed area in different paths will experience different phase shifts produced by the different path length. Figure 5.7 shows a waveguide on a substrate which has a spherical depression on its surface; R is the radius of curvature of the surface depression and 2θ is the vertex angle subtended by the arc of depression. It has been shown that a guided wave propagating in the z direction through the depression will have an additional quadratic phase variation $\exp\left(+jn_{\mathrm{eff},m}k\dfrac{y^2}{2f}\right)$, where

$$f = \frac{R\sin\theta}{2(1-\cos\theta)}. \tag{5.21}$$

Therefore, a plane guided wave will be focused at a distance f after the lens.[16] Vice versa, a cylindrical guided wave originated at distance f before the lens will be collimated. This is known as the geodesic lens. Since all spherical lenses have spherical aberrations, research has been conducted to use an aspheric rotationally symmetric depression to correct the spherical aberration [8]. A numerically controlled, precision lathe has been used for diamond turning the required surface contour on a y-cut LiNbO$_3$ substrate, followed by Ti-diffusion, to make a geodesic lens on a LiNbO$_3$ waveguide with $f = 2$ cm [8].

(C) The Fresnel diffraction lens

In Luneberg and geodesic lenses, the argument in the expression of the quadratic phase shift for a lens, $\exp\left[+jn_{\mathrm{eff},m}k\left(\dfrac{y^2}{2f}\right)\right]$, exceeds multiples of 2π as $|y|$ increases. It is well

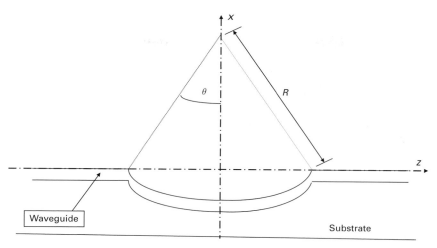

Fig. 5.7.   Cross-sectional view of a geodesic lens.

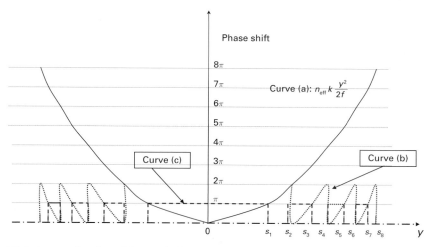

Fig. 5.8.   Digital approximation of the quadratic phase shift – the Fresnel lens. (a) Quadratic phase shift of an ideal lens. (b) Phase shift of an analog Fresnel lens. (c) Phase shift of a digital Fresnel lens.

known that a phase shift of $2n\pi$ is identical to a 0 phase shift. Curve (a) in Fig. 5.8 shows the normal quadratic phase shift for a lens. Curve (b) shows only the value of the phase shift that exceeded $2n\pi$. Clearly, the multiplication of the amplitude and phase of a guided wave as a function of y by a phase shift shown in either (a) or (b) has the same effect. In other words, a component that provides the phase shift shown in (b) will also serve as a lens with a focal length f. Figure 5.8(c) shows a digitalized approximation of (b) in which any phase shift from 0 to π is approximated by π, and any phase shift from π to 2π is approximated by 0. The zones in which the digital change of phase shift is applied to an incoming guided wave are called the Fresnel zones. If we name s_n the $|y|$ at which $n_{\text{eff},m}k\left(y^2/2f\right) = n\pi$ then

$$s_n = \sqrt{\frac{2n\pi f}{n_{\text{eff},m}k}} = \sqrt{n\lambda_m f},$$ (5.22)

where λ_m is the wavelength of the mth order guided-wave mode. The digitalized change of phase for an incident planar TE_0 guided wave has been obtained by depositing rectangular pads of high index materials with length L and in the zone pattern on a planar waveguide [9]. The focusing effect of such a lens could also be viewed as the diffraction effect of the zone pads. Thus it is also known as a Fresnel diffraction lens.

The Fresnel lens is much thinner than the Luneberg or the geodesic lens. However, for large angle oblique incident or divergent waves, the zone structure gives a phase shift distorted from that prescribed in curve (c) of Fig. 5.8.

5.3   Diffraction devices

5.3.1   Grating reflectors and filters

The diffraction structures discussed in the preceding sections have many applications. The most common application is the distributed Bragg reflector (DBR). It is simply a reflection filter as discussed in Section 5.2.1.1(A). *A DBR mirror gives high reflectivity within a specific wavelength range. It has been used most often in semiconductor lasers to replace a cleaved mirror.* Oscillation of unwanted modes in lasers is eliminated by using the wavelength selectivity. The center reflection wavelength and the bandwidth can be tuned by the grating periodicity, the length and the overlap integral between the $\Delta\varepsilon$ and the guided-wave mode (see Section 5.2.1.1). *A modification of a DBR mirror is the use of a grating to achieve distributed feed back (DFB) semiconductor laser oscillation.* Since semiconductor materials have high refractive index, the grating periodicity required for reflection at the first order of the grating is very small. Therefore, the 3rd order of a grating is often used for DBR and DFB. Either DBR or DFB is applicable to channel waveguides as well as to planar waveguides. *Another application is to use the selectivity of such a grating on the effective index to reflect back a specific propagating mode (or a set of modes) in a multimode waveguide.*

In principle, when a voltage is applied to a periodic electrode, its periodic electric field would create a periodic change of index in an electro-optic medium. However, when the periodicity is very small it is almost impossible to make. Its $\Delta\varepsilon$ pattern will have very small thickness t compared to the field pattern of the guided-wave modes, making the electro-optic grating reflector ineffective.

5.3.2   Grating deflector/switch

The Bragg condition for phased matched interaction given in Eq. (5.17) determines the relation between the directions of the incident and deflected waves. Planar guided-wave modes of the same order have the same effective index. Therefore in order to deflect most effectively a guided wave in one direction of propagation to another direction of

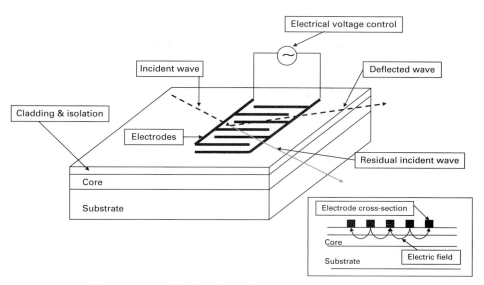

Fig. 5.9. An electro-optic deflection switch. The inset shows the alternating electric field under the electrodes.

propagation as shown in Fig. 5.6, the $\underline{K}$ vector of the grating is perpendicular to the line bisecting the $\underline{\beta}$ of the incident and deflected waves shown in the inset. Equation (5.18) showed that the efficiency of deflection is proportional to $\sin^2(CW)$ where W is the interaction distance given by the length of the grating grooves, and C is the coupling coefficient which is the overlap integral of the grating $\Delta\varepsilon$ with the guided-wave mode shown in Eq. (5.19). For a fixed grating, the primary application would be a guided-wave beam splitter, which deflects a desired fraction of power into another direction.

Note that, for a small $\theta_i = \theta_d$, the T required for a deflection grating is much larger than the T required for a DBR. For such a large T, it is practical to obtain an effective electro-optic $\Delta\varepsilon$ through a voltage applied to a periodic electrode such as the interdigital electrode illustrated in Fig. 5.9. In such an electrode, the applied electric field penetrates into the waveguide below as shown in the inset. The electric field pattern can be calculated by analytical methods such as those discussed in Section 4.3. When the waveguide material is electro-optic, the applied field creates an alternating $\Delta\varepsilon$ between adjacent electrode fingers. In other words, we would then obtain an electro-optic $\Delta\varepsilon$. *The primary objective of an electro-optic deflector is to switch a fraction (or the entire amount) of power into another direction of propagation. The fraction of the power switched into the fixed deflected direction will depend on the applied voltage.*

Three properties of the grating deflector switch should be noted.

(a) The coupling coefficient C in Eq. (5.19) is only moderately dependent on optical frequency ω and the effective index of the mode. The $\underline{K}$ is fixed by the electrode pattern. Even when the Bragg condition in Eq. (5.17) is only satisfied approximately, deflection could still take place.[17] Therefore, the advantage of the grating deflector is that it has a moderate tolerance to incident and deflected angle alignment, mode order and optical wavelength.

(b) There is a serious limitation on the switching speed due to three factors. (1) There is a finite transit time ($\sim Wn_{\mathrm{eff}}/c$) required for the guided waves to travel and interact through the $\Delta\varepsilon$ region. The transit time needs to be much smaller than the switching time. (2) More importantly, the applied voltage creates a uniform pattern of $\Delta\varepsilon$ over the entire electrode area only when the electrode behaves electrically like a lumped element circuit element. For a wide and long interdigital electrode that covers the width of the guided-wave beam (i.e. D) and that provides a reasonably large deflection efficiency, its lumped circuit representation is a large capacitance C. The speed with which the deflector could be driven effectively by a time varying source is limited by the RC time constant, as discussed in Section 4.1.2. (3) It is not possible to obtain traveling wave interaction between the optical waves and electric waves as we discussed in Section 4.4.

(c) In addition to the electro-optic $\Delta\varepsilon$, the metallic electrode pattern may produce a fixed periodic perturbation of the guided waves in the absence of applied voltage. In that case there will be a residual deflected beam. Use of a low index buffer layer between the metal and the guided-wave mode will significantly reduce the residual deflection.

5.3.3   The grating mode converter/coupler

Section 5.2.1.1(B) showed clearly that a co-linear grating with periodicity K satisfying Eq. (5.13) will couple the power between the mth and the nth order planar guided waves. In the reflection mode, the mutual interaction can be utilized as a reflection filter as we have discussed in Section 5.3.1. Since different order modes have different propagation characteristics outside the coupling region such as producing different m-lines through the output prism coupler, the transfer of power by a grating coupler between forward nth and mth order modes can be utilized as a power divider between modes.

5.4   The Star coupler

Diffraction of planar guided waves from an aperture can be utilized to distribute the incident power in a guided wave into a broad range of direction of propagation. An example of such a device is a planar waveguide Star coupler as shown in Fig. 5.10. The power from a given input port (i.e. aperture) fed from a channel waveguide is distributed equally to N output ports. There are N such input ports. Therefore it is an $N \times N$ distributor. It is used in wavelength division multiplexed (WDM) fiber optical systems.

   The Star coupler consists of two arrays of N uniformly spaced identical ports fed from channel waveguides. Each port has width a. Ports (i.e. ends of channel waveguides) in each array are located on a circular arc with radius R. There are two circular arcs facing each other. The center of the circle of the array on the left is at O′ which is also the middle of the circular arc for the array on the right. Vice versa, the center of the circle on the right is at O which is also the middle of the circular arc for the array on the left. The center position of the kth port on the left arc is given by $R\theta_{o,k}$, and the center position of the jth port on the right arc is given by $R\theta'_{o,j}$. The region between the two arrays is a single mode

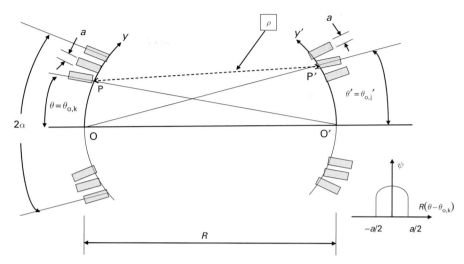

Fig. 5.10.   The Star coupler. (Taken from ref. 10 with permission from IEEE.)

planar waveguide. The power entering the single mode planar waveguide region (from any one of the $2N$ waveguides) will be diffracted and propagated in the yz plane as the generalized guided wave of the planar waveguide. Waveguides on the opposite circular arc are excited by the radiation carried by this generalized guided wave.

The objective of the Star coupler is to maximize the power transfer between any one of the channel waveguides in the left array and any one of the waveguides in the right array. Ideally, there is no power loss and the power from any input waveguide is divided uniformly into the N output channels. In that case the transfer efficiency will be $1/N$. However, this is impossible to achieve in practice. In this section we will analyze the Star coupler using the generalized planar TE_0 guided-wave mode. In particular we will calculate the field at the output array produced by the radiation from a given channel waveguide in the input array. We will calculate the excitation of the mode of the channel waveguide in the output array by this field, thereby determining the power transfer from the input channel to the output channel.

The incident field at each port is the mode of the channel waveguide. Let us assume here that the E_y of the guided-wave mode for all input and output channels in the yz plane is $\psi(y)$ or $\psi(y')$, where y (or y') is the coordinate along the left (or right) circular arc as shown in Fig. 5.10. Transmission between any two ports (i.e. the P port on the left circular arc centered about $\theta_{o,k}$ and the P$'$ port on the right circular arc centered about $\theta'_{o,j}$), is determined by (1) calculating the generalized planar guided-wave field at $y'=R\theta'$ diffracted from P, and (2) calculating the coupling of that field into P$'$.

In order to calculate the field radiated from P to $R\theta'$, we note that the distance between y and y' in the first order approximation of the binomial expansion is

$$\rho = \sqrt{[R\cos\theta' - (R - R\cos\theta)]^2 + (R\sin\theta' - R\sin\theta)^2}$$
$$\cong R - R\sin\theta'\sin\theta \cong R - R\theta\theta'.$$

Thus, for large $\beta\rho$, the field produced by P at P' is

$$E_y(R\theta') \cong \sqrt{\frac{n_{\text{eff}}k}{j2\pi R}} e^{-jn_{\text{eff}}kR} \int_{\theta_{\text{o},k}-\frac{a}{2R}}^{\theta_{\text{o},k}+\frac{a}{2R}} \psi(R\theta) e^{+j2\pi\left(\frac{n_{\text{eff}}}{\lambda}\theta'\right)R\theta} R \, d\theta. \tag{5.23}$$

Here, we have assumed that the field for the kth port is confined approximately within the waveguide as shown in the inset of Fig. 5.10. Note that the phase factor, $-jn_{\text{eff}}kR$, is now a constant on the circular arc on the right. Thus the positioning of the ports on confocal circular arcs serves the function of creating this constant phase factor, similarly to the spherical reflectors in a confocal resonator in three dimensions. The relation between $E_y(R\theta')$ and $\psi(R\theta)$ is related by an integral resembling a Fourier Transform as follows. Using a change of variable, $u = \dfrac{2R}{a}(\theta - \theta_{\text{o},k})$, we obtain

$$E_y(R\theta') \cong a\sqrt{\frac{n_{\text{eff}}}{j\lambda R}} e^{-jn_{\text{eff}}kR} e^{+j2\pi\frac{n_{\text{eff}}R\theta_{\text{o},k}\theta'}{\lambda}} \phi(R\theta'),$$

where

$$\phi(R\theta') = \frac{1}{2}\int_{-1}^{+1} \psi\left(\frac{au}{2}\right) e^{+j2\pi\left(\frac{n_{\text{eff}}a\theta'}{2\lambda}\right)u} \, du. \tag{5.24}$$

Since the $\psi(au/2)$ is identical for all the waveguides, the ϕ factor is independent of $\theta_{\text{o},k}$. The E_y is only dependent on the center position $R\theta_{\text{o},k}$ of the input channel through the factor $\exp\left(j2\pi\dfrac{n_{\text{eff}}R\theta_{\text{o},k}\theta'}{\lambda}\right)$. Let the total E_y at $R\theta'$ be expressed as a summation of the fields of all the channel guides, $\psi_i(R\theta')$, on the right circular arc array plus the stray guided-wave fields in the gaps between channel guides, $\zeta(R\theta')$. Let us assume, as an approximation, that there is negligible overlap among all the ψ_i and the ζ. Then

$$E_y(R\theta') = \sum_i b_i\psi_i(R\theta') + \zeta(R\theta'). \tag{5.25}$$

Here, $\psi_i(R\theta')$ is the ψ centered about $\theta_{\text{o},i}$. Multiplying both sides by $\psi_j^*(R\theta')$ and integrating with respect to $R\theta'$ from $-\infty$ to $+\infty$, we obtain

$$\int_{\theta_{\text{o},j}-\frac{a}{2R}}^{\theta_{\text{o},j}+\frac{a}{2R}} E_y(R\theta')\psi(R\theta')R \, d\theta' \cong b_j \int_{\theta_{\text{o},j}-\frac{a}{2R}}^{\theta_{\text{o},j}+\frac{a}{2R}} |\psi(R\theta')|^2 R \, d\theta'.$$

Utilizing once more the change of variable, $u' = \dfrac{2R}{a}(\theta' - \theta'_{\text{o}})$, we obtain

$$|b_j|^2 \left[\frac{a}{2}\int_{-1}^{+1} \left|\psi\left(\frac{a}{2}u' + R\theta_{\text{o},j}\right)\right|^2 du'\right]^2 = \left[\frac{n_{\text{eff}}a^4}{\lambda R}|\phi(R\theta_{\text{o},k})|^2|\phi(R\theta_{\text{o},j})|^2\right], \tag{5.26}$$

or

$$|b_j|^2 = \frac{4n_{\text{eff}}a^2}{\lambda R} \frac{\left|\phi\left(R\theta_{\text{o},k}\right)\right|^2 \left|\phi\left(R\theta_{\text{o},j}\right)\right|^2}{\left[\displaystyle\int\limits_{-1}^{+1} \left|\psi\left(\frac{a}{2}u + R\theta_{\text{o},k}\right)\right|^2 \mathrm{d}u\right]^2}. \tag{5.27}$$

Since the power contained in the total E_y is proportional to $\int |E_y|^2 R\,\mathrm{d}\theta$ which is approximately equal to $\sum_i |b_i|^2 \int |\psi|^2 R\,\mathrm{d}\theta$, $|b_j|^2$ is the power transfer from the channel waveguide centered at $\theta_{\text{o},k}$ to the channel waveguide centered at $\theta_{\text{o},j}$.

In an actual Star coupler, R, N and "a" are designed to optimize the power transfer. Dragone and his colleagues have optimized the design which gives $0.34(1/N)$ to $0.55\,(1/N)$ of the input power to any one of the output channels [10].

5.5 The acousto-optical scanner, spectrum analyzer, and frequency shifter

An acousto-optical deflector (or scanner) is a device that deflects a planar guided-wave mode into a different direction by a grating generated from a surface acoustic wave (SAW). The surface acoustic wave is generated from an electric signal applied to a SAW transducer such as the interdigital transducer illustrated in Fig. 3.11. The strain from the acoustic wave creates a surface layer of traveling refractive-index waves with periodic index variation. An approximate mathematical expression of a traveling grating generated from a CW single frequency acoustic wave is presented below in Eq. (5.28). The periodicity of the refractive-index wave is determined by the SAW wavelength. The generation of the traveling refractive-index wave has been discussed in Section 3.4.

When there is an incident optical guided wave and when the phase matching condition along both the lateral and the longitudinal directions is satisfied, efficient diffraction occurs. Optical energy in the incident wave is transferred to the deflected wave that has a slightly different direction of propagation than the incident wave as illustrated in Fig. 3.12. The direction of the deflected wave depends on the acoustic wavelength which is the acoustic velocity divided by acoustic frequency. At the same time, the optical frequency of the deflected wave is shifted slightly from that of the incident wave. Acousto-optical interaction under the phase matched condition has already been analyzed and discussed in Section 3.5.3. Figure 5.11 illustrates an integrated optical spectrum analyzer [8].

In the acousto-optical spectrum analyzer illustrated in Fig. 5.11, the single mode LiNbO$_3$ waveguide has polished vertical input and output ends. An incident beam of guided wave in the 0th order TE mode is excited by the semiconductor laser butt coupled to the waveguide. The small lateral size of the oscillating mode of the semiconductor laser produces a divergent optical planar guided-wave beam as we have discussed in Section 5.1.1. A geodesic lens as described in Section 5.2.2.2(B) is used to collimate the divergent beam. A surface acoustic wave (SAW) transducer generates a surface acoustic wave in the y direction. It is W wide in the z direction. The SAW is eventually absorbed by

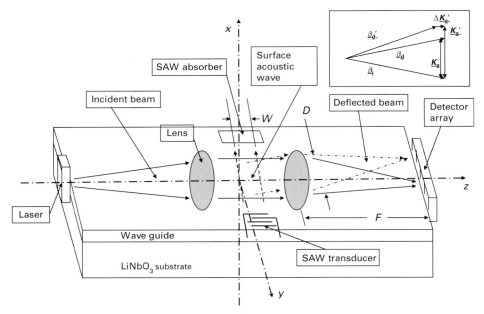

Fig. 5.11.   The acousto-optical spectrum analyzer. The inset shows the matching of $\underline{K}_a$, $\underline{\beta}_i$ and $\underline{\beta}_d$ as the acoustic $\underline{K}_a$ is varied.

the acoustic absorber. Hence there is no reflected SAW. When the SAW transducer is driven by a RF signal, the SAW generates a traveling wave $\Delta\varepsilon$ grating. For a CW RF signal applied to the SAW transducer, the $\Delta\varepsilon$ is described mathematically in Eq. (3.28) as

$$\Delta\varepsilon(x,y,z,t) = \Delta\varepsilon \cos\left(\underline{K}_a \cdot \underline{\rho} - \Omega t\right) \mathrm{rect}\left(\frac{t/2+x}{t/2}\right) \mathrm{rect}\left(\frac{W/2-z}{W/2}\right),$$

$$\underline{K}_a = K_a \underline{i}_y, \quad \underline{\rho} = y\underline{i}_y + z\underline{i}_z, \quad \varepsilon = n^2\varepsilon_0, \quad \Delta\varepsilon = (2n\Delta n)\varepsilon_0. \tag{5.28}$$

Note that K_a is proportional to acoustic frequency Ω, $K_a = 2\pi/\lambda_a = \Omega/v_{ac}$. The $\Delta\varepsilon(x,y,z,t)$ is W wide. Although the magnitude of the SAW has an x variation, the magnitude of $\Delta\varepsilon(x,y,z,t)$ in the x direction is approximated by a uniform value of $\Delta\varepsilon$ from $x=0$ to $x=-t$ in Eq. (5.28). The Bragg condition for phase matched interaction is given in Eq. (3.45),

$$\left(\underline{\beta}_i - \underline{\beta}_d\right) \cdot \underline{\rho} = \mp \underline{K}_a \cdot \underline{\rho} \quad \text{or} \quad \underline{\beta}_d = \underline{\beta}_i \pm \underline{K}_a. \tag{5.29}$$

When the Bragg condition is satisfied, the $\Delta\varepsilon$ grating deflects part of the optical power in the incident collimated beam in the θ_i direction into a deflected collimated beam in the θ_d direction. Both the incident and the deflected beam are focused by a second geodesic lens on to the output edge of the LiNbO$_3$ waveguide. Since θ_i and θ_d are different, the incident and the deflected beams will have different focused spots along the y direction. According to Eq. (3.48) the optical power of the deflected beam will be proportional to $\sin^2(C_a W)$. According to Eq. (3.44), (3.30), and (3.28), C_a is proportional to $\sqrt{P_{ac}}$, where P_{ac} is the acoustic power provided by the RF signal. As the acoustic frequency changes, the direction of the collimated deflected beam will change, resulting in a change of the position of the focused spot. In the spectrum analyzer, a detector array is used to detect the optical power coupled out at different y positions. Therefore, change in the detected position of the focused radiation output is a measure of the RF frequency, while the amount of the optical power in the deflected spot is a measure of the RF power applied to the SAW transducer.

Note also that, according to the analysis presented in Section 3.5.3, the optical radiation in the deflected beam is related to the optical frequency of the incident beam by $\omega_d = \omega_i \mp \Omega$. The shift in ω_d from ω_i is used to obtain frequency shifting. When the detector array is removed from the spectrum analyzer shown in Fig. 5.11 the component becomes an acousto-optical scanner.

The acoustic transducer, the lenses and the position of the optical input of the incident guided wave cannot be repositioned after the devices have been made. Thus the directions of the incident collimated optical guided wave and the acoustic wave are fixed. Since the periodicity of the acousto-optical grating is determined by the frequency of the electrical signal applied to the SAW transducer, the Bragg condition is strictly satisfied for a given θ_i only at the center frequency Ω. When the acoustic (i.e. RF) frequency is varied, the relation between $\underline{\beta}_i, \underline{\beta}_d$ and $\underline{K}_a$ will deviate from the Bragg condition. The diffraction efficiency will decline when the deviation from the Bragg condition becomes bigger.

We can evaluate again the effect of mismatch between $\underline{\beta}_i, \underline{\beta}_d$ and $\underline{K}_a$ by coupled mode analysis.[18] When the acoustic frequency shifts from Ω to Ω', K_a changes to K'_a. The optical frequency of the diffracted wave is $\omega'_d = \omega_i - \Omega'$. Note that $\underline{\beta}'_d$ will now be oriented in a new direction θ'_d to satisfy the phase matching condition in the y direction. Other waves with $\underline{\beta}$s that do not meet the phase matching condition in the y direction will have negligible amplitude. The shift of θ'_d as a function of K'_a is the principal mechanism for controlling the direction of deflection. $\underline{\beta}_i, \underline{K}'_a$ and $\underline{\beta}'_d$ are illustrated in the inset of Fig. 5.11.

The relation between $\underline{\beta}_i, \underline{\beta}'_d$ and $\underline{K}'_a$ can now be stated as

$$\underline{\beta}'_d = \underline{\beta}_i + \underline{K}'_a + \Delta\underline{K}'_a, \quad \text{or} \quad \Delta\underline{K}'_a = \Delta K'_a \underline{i}_z = n_{\text{eff}}k\left(\cos\theta'_d - \cos\theta_i\right)\underline{i}_z. \tag{5.30}$$

When we include the $\Delta K'_a$, the coupled mode equation (3.44) becomes

$$\frac{da_i}{dz} = -ja'_d C_a e^{j\Delta K'_a z}, \quad \text{and} \quad \frac{da'_d}{dz} = -ja_i C_a e^{-j\Delta K'_a z}. \tag{5.31}$$

The solution for $|a'_d| = 0$ at $z = 0$ is

$$a_i = A e^{\frac{j\Delta K'_a}{2}z}\left[\cos\sqrt{C_a^2 + \left(\frac{\Delta K'_a}{2}\right)^2}\,z - j\frac{\left(\frac{\Delta K'_a}{2}\right)}{\sqrt{C_a^2 + \left(\frac{\Delta K'_a}{2}\right)^2}}\sin\sqrt{C_a^2 + \left(\frac{\Delta K'_a}{2}\right)^2}\,z\right],$$

$$a'_d = -\frac{-jC_a A}{\sqrt{C_a^2 + \left(\frac{\Delta K'_a}{2}\right)^2}}e^{-j\frac{\Delta K'_a}{2}z}\left[\sin\sqrt{C_a^2 + \left(\frac{\Delta K'_a}{2}\right)^2}\,z\right]. \tag{5.32}$$

The degradation of the deflection efficiency from a_i into a'_d is small when $\dfrac{\Delta K'_a}{2} \ll C_a$.

Acousto-optic deflection has a number of applications. (1) When the acoustic frequency is scanned, the direction of the deflected wave changes. Thus an acousto-optical deflector can be used as an optical scanner. (2) The optical frequency of the deflected beam is shifted from the frequency of the incident optical beam. Thus an acousto-optical deflector is used sometimes as a frequency shifter. (3) When the acoustic signal has a complex RF frequency spectrum, the optical energy deflected into various directions can be used to measure the power contained in various RF frequency components, known as the acousto-optical RF spectrum analyzer. How we utilize acousto-optical deflection for spectral analysis of RF signals, for scanning of an optical beam, or for slightly shifting the optical frequency varies. Applications using these devices have different objectives. Their electrical and optical performance is discussed separately in the following sections.

5.5.1   The optical scanner

The physical structure of an optical scanner will be similar to that of the spectrum analyzer shown in Fig. 5.11 without the detector array. The CW SAW at frequency Ω is used to transfer the power in the collimated incident optical guided-wave into a deflected wave. The deflected spot is scanned as Ω is scanned. The focused spot of the deflected wave is the output. Each spot has a finite size. In order to resolve different spots, any change in focused position must be larger than the spot size. In any scanner, we like to obtain the largest number of resolvable spots. It is also desirable to transfer uniformly the maximum amount of optical power in the incident beam into the output ports of the deflected beam. The speed at which the scanning can be achieved is also important.

From the above discussion, we can make the following conclusions.

(A) The wider the range of $\Delta\Omega$ that we scan, the larger is the scanned $\Delta\theta_d$. The range within which $\Delta\theta_d$ can be scanned depends on two factors. (1) The bandwidth of the transducer. (2) The frequency dependence of the attenuation of the SAW. In Eq. (5.28), we have assumed that the SAW is not attenuated. In reality the attenuation of a SAW increases with Ω. It becomes very high at the GHz range.

(B) For a given lens with focal length F, the distance that the focused spot moves for a given $\Delta\theta_d$ is $F\Delta\theta_d$. The minimum diffraction-limited focused spot size of an ideal lens is of the order of $F\lambda_g/D$, where D is the width of the collimated guided wave, and $\lambda_g = \lambda/n_{\text{eff},m}$ is the wavelength of the guided wave. Therefore the number of resolvable spots that can be scanned is $\Delta\theta_d D/\lambda_g$.

(C) There are limitations on how wide a D can be employed. (1) The D is limited by the available size of the chip. (2) The D may be limited by the desired speed to switch from one Ω to another. The time required to swing from one Ω to another must be longer than the transit time for the SAW to cover the width of the optical guided wave D, i.e. D/v_{ac}. The larger the D, the slower is the switching time from one spot to another.

(D) In addition, in order to transfer optical power efficiently into the deflected spot we like to have $C_a W \approx \pi/2$. However, in order to generate SAW at higher frequencies the spacing of the interdigital electrodes which controls the height of the acoustic wave, needs to be reduced. The higher the SAW frequency, the smaller is the height t of the acoustic wave. The smaller the t, the smaller is the coupling coefficient C_a. The reduction of C_a and increase of ΔK_a limits the uniformity of the power deflected into the spots.

(E) Obviously, the performance of the scanner will also be affected by the design of the SAW transducer which has not been covered in this book.

5.5.2   The acousto-optical RF spectrum analyzer

Note that, for small $\sqrt{C_a^2 + (\Delta K_a'/2)^2}\, W$ and for $\Delta K_a' \ll C_a$, the intensity of the diffracted beam, $|a_d'|^2$, is proportional to $(C_a W)^2$ which is proportional to P_{ac}. When the SAW contains many frequency components, it means that the intensity of the deflected beams at a given θ_d' will be proportional to the power of the acoustic surface wave component at the frequency Ω'. In other words, when one measures the optical power deflected into different directions θ_d', the detected optical power measures the RF power at the frequency Ω' applied to the transducer. The performance of an acousto-optical spectrum analyzer will be measured by the frequency resolution at which the spectral components can be distinguished and by the dynamic range at which the RF power of different frequency components can be measured. Note that the transit time for the acoustic wave to travel across the optical beam is D/v_{ac}. Therefore any variation of the RF spectrum within a time period shorter than the transit time will yield errors in acousto-optical spectrum analysis.

The diffraction limited spot size of an ideal lens with focal length F is $F\lambda_g/D$ where D is the size of the incident beam. Vice versa, any incident beam within an angular spread λ_g/D will be focused within the spot. According to Eq. (5.30),

$$\Delta\theta_d' = \frac{(\Delta\Omega/2\pi)\lambda_g}{v_{ac}}, \tag{5.33}$$

using the small angle approximation of θs. Therefore, the minimum measurable spread of acoustic frequency corresponding to the diffraction limited angular spread $\Delta\theta_d' = \lambda_g/D$ is

$$\Delta f_{ac} = \frac{\Delta\Omega}{2\pi} = \frac{v_{ac}}{D}. \tag{5.34}$$

If the detector array has detector spacing s_d which is just larger than the diffraction limited spot size, the spread of acoustic frequency corresponding to a single detector cell is

$$\Delta f_{ac} = \frac{v_{ac}}{D} \frac{s_d}{(F\lambda_g/D)} = \frac{v_{ac}s_d}{F\lambda_g}, \tag{5.35}$$

where Δf_{ac} is the minimum frequency resolution of the spectrum analyzer. For LiNbO$_3$, using $v_{ac} = 3500$ m/s, $n_{eff} = 2.2$, $s_d = 10\ \mu$m, and $F = 45$ mm, Δf_{ac} is about 2 MHz.

The dynamic range of a single spectral component in the spectrum analyzer will be determined by the logarithm of the ratio of the maximum signal power to the minimum detectable signal power in dB. The minimum detectable signal power at a given frequency Ω is the signal power that produces a photo-current in the detector equal to the photo-current contributed from all noise sources. These sources include the detector shot and thermal noise, the laser noise, numerous mechanisms that couple the stray optical power from the incident beam, the signal component, as well as adjacent strong spectral components entering the detector through mechanisms such as radiation from side diffraction lobes, in-plane scattering, etc. [8]. The maximum detectable power is the maximum allowed signal power without producing large spurious signals due to non-linearity. In reality the maximum signal power is often limited by the available laser power.

5.5.3   The acousto-optical frequency shifter

In many respect, the acousto-optical frequency shifter is a very simple device. The SAW transducer needs only to generate efficiently a SAW at the desired shift of frequency Ω. The magnitude of the SAW needs to be controlled such that $C_a W \approx \pi/2$. The focused spot size should match the mode of the output fiber so that the frequency shifted optical power is transferred efficiently into the output channel, e.g., an optical fiber.

Notes

1. For waveguides that have an inclined end surface, the incident beam needs to be directed at an angle according to Snell's law of refraction.
2. Snell's law of refraction applies to the directions of propagation of the incident beam and the excited guided-wave beam where the indices are the index of the air and the effective index of the guided wave.
3. Note that selective excitation of a higher order mode requires a phase variation of the input radiation to match that of the guided-wave mode. This is difficult to do.
4. The transmitted wave is in phase with the plane wave representing the mth order waveguide mode which has been excited by other optical rays at all prior positions of z because $n_p \cos \theta_p m = n_{\text{eff},m}$. If $n_p \cos \theta_p \neq n_{\text{eff},m}$, the transmitted wave will have varying phase relation with respect to the plane waves of the mth order mode excited at different z positions, resulting in cancellation of the accumulated guided wave.
5. Similarly to the tunneling of electrons between energy gaps, the transmission through the evanescent field in the air gap is also called tunneling.
6. For example, LiNbO$_3$ waveguides have $n_{\text{eff},m} \approx 2.2$. Then the evanescent field variation at the 1.5 μm wavelength is $\exp(-8.2 \times 10^6 x)$. This means that the air gap should be $\approx 10^{-7}$ m in order to get effective interaction.
7. The intensity of the guided wave is usually monitored by observing the intensity of the radiation scattered from the waveguide by defects or surface irregularities or by employing a second prism to couple out the guided-wave energy. The energy in the guided-wave mode will appear as an m-line of the output prism. Sometimes the waveguide is terminated by a smooth vertical end surface or a tapered end. In that case, the intensity of the guided-wave mode can be observed or measured directly at the output end.
8. From another point of view, a plane wave in the air is diffracted by the grating into another plane wave in the core of the waveguide. When the Bragg condition is satisfied, the diffracted plane wave, internally reflected at the core boundary, is the guided-wave mode in the core.

9. See Sections 1.2.3.3 and 1.2.4.3 for discussion on the cut-off condition of the modes.
10. An important advantage of using the prism method to measure the waveguide loss is that the input coupling can be held constant while the output coupler is moved. The output coupling efficiency is reproducibly 100%.
11. Equation (2.12) is written for channel waveguide modes. To apply it to planar waveguide modes, we consider the guided wave to be uniform in y and have unity lateral width. When grating width W covers the entire width of the guided wave, the integration in the y direction will yield just unity.
12. See Sections 1.1.1, 1.2.3.5 and 1.2.3.6 for further discussion of equivalence between air and substrate modes and plane waves propagating in the substrate or air.
13. Since the transfer efficiencies to different modes depend on the Fourier coefficient of the expansion, i.e. the shape of the dielectric grating, a grating with a blazed groove may be used to reduce diffraction into unwanted orders [7].
14. In comparison with Eq. (3.45), we have assumed here that $n_{\mathrm{eff},m} \cos \theta_{\mathrm{i}} \cong n_{\mathrm{eff},m} \cos \theta_{\mathrm{d}}$. This means that effective deflection occurs only when $\theta_{\mathrm{i}} \cong \theta_{\mathrm{d}}$.
15. Since $n_{\mathrm{eff},m}$ is close to $n'_{\mathrm{eff},m}$, the reflection will usually be very small. However when $n_{\mathrm{eff},m} \cos \theta_{\mathrm{i}} > n'_{\mathrm{eff},m}$ total reflection will take place at the boundary.
16. Note that f is independent of effective index or wavelength. It depends only on the geometry.
17. Deflection slightly off the Bragg angle will be discussed for the acousto-optic grating in Section 5.4.
18. Coupled mode analysis is applicable because the incident and deflected beams overlap each other in the region of the yz plane where the SAW exists.

References

1. W. S. C. Chang, *Principles of Lasers and Optics*, Section 1.3, Cambridge University Press (2005).
2. P. M. Morse and H. Feshbach, *Methods of Theoretical Physics*, Chapter 7, McGraw-Hill (1953).
3. P. K. Tien and R. Ulrich, Theory of prism-film coupler and thin-film light guides. *J. Opt. Soc. Am.*, **60** (1970) 1325.
4. D. L. Lee, *Electromagnetic Principles of Integrated Optics*, John Wiley and Sons (1986).
5. J. H. Harris, R. K. Winn, and D. G. Dalgoutte, Theory and design of periodic couplers. *Appl Opt.*, **11** (1972) 2234.
6. W. S. C. Chang, M. W. Muller, and F. J. Rosenbaum, Integrated optics. In *Laser Applications*, Vol. 2, ed. M. Ross, Academic Press (1972).
7. T. Tamir, Beam and waveguide couplers. In *Integrated Optics*, 2nd ed., *Topics in Applied Physics*, Vol. t, Springer (1979).
8. M. C. Hamilton and A. E. Spezio, Spectrum analysis with integrated optics. Chapter 7 in *Guided Wave Acousto-optics*, ed. Chen S. Tsai, Springer (1990).
9. P. R. Ashley, Fresnel lens in a thin film waveguide. *Appl. Phys. Lett.*, **33** (1978) 490.
10. C. Dragone, Efficient NxN coupler using Fourier optics. *J. Lightwave Tech.*, **7** (1989) 479.

6 Channel waveguide components

Fields in channel waveguides are confined to the vicinity of the core within a few μm in both the lateral and the depth directions. There are two main advantages of the lateral confinement of channel guided-wave modes:

(1) The RF electric field required to obtain an electro-optical effect such as electro-optic change of index or electro-absorption needs only to exist in a small region around the core. The required electric field in a small region can be achieved with just a moderate RF voltage applied to the electrodes. Furthermore, when the electrodes are fabricated parallel to the channel waveguides, the electro-optical change of index or electro-absorption produced by a propagating RF signal can be synchronized with the propagation of the guided wave in a traveling wave interaction as discussed in Chapter 4. Thus the electro-optical modulation at high frequencies may be carried out effectively.[1]

(2) Most optoelectronic devices are eventually connected to single mode optical fibers. The optical field pattern of the channel waveguides can be designed such that it matches well with the field pattern of single mode optical fibers or tapered fibers, providing high efficiency transmission of optical power to and from the low loss single mode fibers.

Traditionally, guided-wave devices have been discussed in the literature according to the type of optical interactions they utilize, such as directional coupling or electro-absorption. However, for a given application such as switching, there are many competing channel waveguide optoelectronic devices that may utilize different types of interaction and electro-optical effects to achieve the desired operational function. The choice of the device needs to be based on its combined optical and electrical characteristics best suited for that application. Therefore the discussion of the devices in this chapter is presented according to the different optoelectronic functions to be performed.

6.1   Passive waveguide components

In many applications, passive waveguide devices are used to split or distribute optical power into different waveguides, to filter the optical radiation according to its wavelength or to provide true time delay of pulsed radiation. Besides the optical properties such as insertion loss, power distribution and wavelength bandwidth, the performances of these devices are characterized by their ability to interface effectively with single mode optical fibers. There is no RF electrical signal controlling their characteristics.

6.1.1 The power divider

In guided-wave and fiber optical systems, power dividers are needed to channel specific fractions of input power into different output channels. The power dividers may also be interconnected to other waveguide devices. Eventually, single mode fibers are connected to the waveguide devices. The performance of power dividers is measured by their output power distribution, wavelength response, insertion loss, physical size and coupling loss to other components or fibers. The matching of the channel waveguide mode and the fiber mode is important to minimize the coupling loss.[2]

6.1.1.1 The Y-branch equal-power splitter

An adiabatic symmetric single mode Y-branch waveguide splits the optical power in the input waveguide equally into two output waveguides. In this device, a single mode waveguide is interconnected with two identical single mode waveguides in a Y-branch configuration. The device was discussed in Section 2.3.4.2 and illustrated in Fig. 2.7(a). The waveguide are coupled into other waveguide devices or optical fibers. The performance of any splitter is defined by its insertion loss and distribution of input power into the two output waveguides. Depending on the waveguide mode pattern, the tapering angle, the sharpness of the intersection, the material and fabrication tolerance, there is an excess insertion loss, produced by the power scattered into the substrate at the intersection. In addition, there are propagation losses and coupling losses to and from the optical fiber. The major cause of scattering loss is often the imperfections created by the fabrication process. Since the scattering increases as the dielectric discontinuity of the imperfections increases, the excess loss is more likely to be higher in a Y-branch that has more abrupt boundaries and a larger difference between the index of the core and the claddings. *Most Y-branches are symmetrical, meaning that the power is divided equally into the two output waveguides.*[3] Although an unsymmetrical Y-branch should, in principle, be able to divide the power unequally into the two output waveguides (see Section 2.3.4.4), its performance is sensitive to variation of Y-branch and waveguide configurations. Instead, the directional coupler discussed in Sections 2.2.4 and 2.3.2 is the preferred method to split power unequally into two waveguides. The performance of the Y-branch equal-power splitter is only very mildly dependent on wavelength because the splitting ratio is independent of the wavelength and the excess and propagation losses vary slowly with wavelength.

6.1.1.2 The directional coupler

A directional coupler consists of two parallel waveguides, A and B, that have a coupling region with length W. Outside the coupling region, the waveguides are well isolated from each other. They function as individual isolated waveguides. Within the coupling region, the guided waves propagating in the two waveguides are coupled to each other via their evanescent fields. The powers of the guided waves in A and B transfer back and forth as they propagate in the coupling region. The power distribution in the two waveguides at the end of the coupling region determines the distribution of the output power. The directional coupler was discussed in Section 2.2.4 and illustrated in Fig. 2.4(a). If the device is used standing alone, a single mode optical fiber (or a waveguide from another device) will be coupled to the input waveguide A and to the output waveguides A and B.

Let the coupling region begin at $z = 0$ and end at $z = W$. Let the amplitude of the guided wave in waveguide A be a_A, and the amplitude of the guided wave in waveguide B be a_B. At $z = 0$, let $a_A = A$ and $a_B = 0$. From Eq. (2.22), we obtain:

$$
a_A = A e^{j\frac{\Delta\beta}{2}W} \left[\cos\left(\sqrt{C_{BA}C_{AB} + (\Delta\beta/2)^2}\, W \right) \right.
$$

$$
\left. -j\frac{(\Delta\beta/2)}{\sqrt{C_{BA}C_{AB} + (\Delta\beta/2)^2}} \sin\left(\sqrt{C_{BA}C_{AB} + (\Delta\beta/2)^2}\, W \right) \right],
$$

$$
a_B = \frac{-jC_{AB}A}{\sqrt{C_{BA}C_{AB} + (\Delta\beta/2)^2}} e^{-j\frac{\Delta\beta}{2}W} \sin\left(\sqrt{C_{BA}C_{AB} + (\Delta\beta/2)^2}\, W \right),
$$

$$
(6.1)
$$

where $\Delta\beta = (n_{eff,A} - n_{eff,B})k$, W is the length, and C_{AB} and C_{BA} are coupling coefficients in Eq. (2.11) for the coupling section. Clearly the amplitude ratio a_B/A is controlled by $\Delta\beta$, C and the length of interaction $z = W$.

In most directional couplers, A and B are identical waveguides, thus $\Delta\beta = 0$ and $C_{AB}=C_{BA}=C$. At $z = W$, when propagation and coupling losses are neglected, 100% power is transferred from waveguide A to waveguide B for $CW = \left(n + \frac{1}{2}\right)\pi$, and all power is retained in waveguide A for $CW=n\pi$. Vice versa, there is 100% transfer of power from waveguide B to waveguide A when $CW = \left(n + \frac{1}{2}\right)\pi$, and all power is retained in waveguide B when $CW = n\pi$. In practice, there will be a small insertion loss caused by propagation loss and the excess scattering loss produced from the fabrication processes.

The directional coupler is a reciprocal device. Reflected optical power in the output waveguides will also be distributed in the same ratio back to the input waveguides. Thus the waveguide at the input that does not have the incident radiation could be used to monitor the reflections from the output. Most commonly, the waveguides at the input end will be match terminated, meaning that the reflected power to the input guides will be radiated or absorbed without further reflection. Since the coupling coefficient C and $\Delta\beta$ are dependent on the wavelength, its performance will have a moderate wavelength dependence. When the coupling coefficient C or the $\Delta\beta$ of the modes is controlled electro-optically, it becomes a directional coupler switch (or modulator) which will be discussed in Section 6.1.2.5. The directional coupler could also easily be made directly from two coupled optical fibers with cladding partially removed to provide the coupling via the evanescent field. The length of this interaction region and the proximity of fiber cores control the power splitting ratio between the two fibers. The advantage of a fiber directional coupler is that there is no need to couple the fiber to the waveguide.

There are other power dividers besides the 3 dB Y-branch and directional couplers. Three examples are discussed here.

6.1.1.3   The Star coupler $1 \times N$ splitter

The Star coupler discussed in Section 5.4 is basically a $1 \times N$ splitter. It is illustrated in Fig. 5.10. In this case, the input power can be incident into any one of the waveguides located on the input circular arc. The output waveguides are located on the opposite circular arc. The region between the input circular set of waveguides and the output circular set of waveguides is a planar waveguide. The diffraction of the electric field of the input waveguide in the planar waveguide distributes the input power to all the output waveguides. Ideally, the input power will be divided equally into all the output wave-guides. In reality, Dragone has demonstrated a design which gives $0.34(1/N)$ to $0.55(1/N)$ to the output waveguides [1].[4]

Alternatively, directional couplers discussed in the previous section could also be used repeatedly to split the input power into more than one output. There are three advantages of the Star coupler.

(1) *The Star coupler is a much simpler device for $N > 2$.*
(2) *The $1 \times N$ splitting can be achieved from any one of the N input guides.*
(3) *There is a well-controlled phase variation of the optical guided waves at each output channel waveguide (see Eq. (6.11)) while the phases of the optical radiation in the outputs of the multiple-directional couplers are very difficult to control. However, the power can only be distributed equally to the output waveguides in a Star coupler, while the power transfer ratio of each of the multiple-directional couplers could be adjusted individually.*

6.1.1.4   The Y-branch variable power distributor

In Section 2.3.4.2(b) we discussed a device which consists of two symmetrical Y-branch couplers back-to-back connected by a two-mode waveguide between the two Y-branches. It is illustrated in Fig. 2.8(a). This is also a power distributor, for whatever the radiation input into the two input waveguides of the input Y-branch, it excites both the symmetric mode $\underline{e}_s$ and anti-symmetric mode $\underline{e}_a$ at the beginning of the two-mode section.

Let the amplitude of the guided-waves in the two isolated identical waveguides A and B at the input end of the Y-branch (i.e. at $z = -L_0$ in Fig. 2.8(a)) be A_{in} and B_{in}. Then the electric field at $z = -L_0$ can be represented either by the modes of the isolated waveguides or by the super modes of the two waveguides combined as follows:

$$A_{in}(z = -L_0)\underline{e}_A + B_{in}(z = -L_0)\underline{e}_B = A_s(z = -L_0)\underline{e}_s + A_a(z = -L_0)\underline{e}_a, \qquad (6.2)$$

where $\underline{e}_A$ and $\underline{e}_B$ are the modes of the isolated waveguides and $\underline{e}_s$ and $\underline{e}_a$ are the symmetric and antis-ymmetric modes of the combination of waveguides A and B. Note that A_s and A_a are amplitudes of the symmetric and anti-symmetric modes. From Section 2.3.2,

$$\underline{e}_s = \frac{1}{\sqrt{2}}(\underline{e}_A + \underline{e}_B), \qquad \underline{e}_a = \frac{1}{\sqrt{2}}(\underline{e}_A - \underline{e}_B).$$

From the orthogonality property of the modes in lossless waveguides, we have

$$
\begin{aligned}
A_s \int \underline{e}_s \bullet \underline{e}_s^* dx\, dy &= \left[A_{in} \int \underline{e}_A \bullet \underline{e}_s^* dx\, dy + B_{in} \int \underline{e}_B \bullet \underline{e}_s^* dx\, dy \right] \\
&= \frac{A_{in}}{\sqrt{2}} \int \underline{e}_A \bullet \underline{e}_A^* dx\, dy + \frac{B_{in}}{\sqrt{2}} \int \underline{e}_B \bullet \underline{e}_B^* dx\, dy
\end{aligned}
\tag{6.3}
$$

$$
\begin{aligned}
A_a \int \underline{e}_a \bullet \underline{e}_a^* dx\, dy &= A_{in} \int \underline{e}_A \bullet \underline{e}_a^* dx\, dy + B_{in} \int \underline{e}_B \bullet \underline{e}_a^* dx\, dy \\
&= \frac{A_{in}}{\sqrt{2}} \int \underline{e}_A \bullet \underline{e}_a^* dx\, dy - \frac{B_{in}}{\sqrt{2}} \int \underline{e}_B \bullet \underline{e}_B^* dx\, dy.
\end{aligned}
\tag{6.4}
$$

For a symmetrical Y-branch,

$$
\int \underline{e}_A \bullet \underline{e}_A^* dx\, dy = \int \underline{e}_B \bullet \underline{e}_B^* dx\, dy = \int \underline{e}_s \bullet \underline{e}_s^* dx\, dy = \int \underline{e}_a \bullet \underline{e}_a^* dx\, dy.
$$

Therefore, we obtain:

$$
A_s = \frac{A_{in} + B_{in}}{\sqrt{2}} \quad \text{and} \quad A_a = \frac{A_{in} - B_{in}}{\sqrt{2}}.
\tag{6.5}
$$

When the symmetric and anti-symmetric modes propagate adiabatically from $z = -L_0$ to $z = +L_0$, their amplitudes and phases at the output are

$$
A_s(z = +L_0) = A_s(z = -L_0)e^{-j\phi_{s,in}}e^{-j\beta_s(2L_c)}e^{-j\phi_{s,out}},
\tag{6.6}
$$

$$
A_a(z = +L_c) = A_a(z = -L_c)e^{-j\phi_{a,in}}e^{-j\beta_a(2L_c)}e^{-j\phi_{a,out}}.
\tag{6.7}
$$

Here, $\phi_{s,in}$ and $\phi_{a,in}$ are the phase shift of the symmetric and anti-symmetric modes at the input Y-branch, $\phi_{s,out}$ and $\phi_{a,out}$ are the phase shift of the symmetric and anti-symmetric modes at the output Y-branch, and β_s and β_a are the propagation wave numbers of the symmetric and anti-symmetric modes in the two-mode waveguides. The two-mode waveguide is $2L_c$ long. At the output we have

$$
\begin{aligned}
A_{out}(z = +L_c) &= \frac{A_{in} + B_{in}}{2} e^{-j\phi_{s,in}}e^{-j\phi_{s,out}}e^{-j2\beta_s L_c} \\
&\quad + \frac{A_{in} - B_{in}}{2} e^{-j\phi_{a,in}}e^{-j\phi_{a,out}}e^{-j2\beta_a L_c},
\end{aligned}
\tag{6.8}
$$

$$
\begin{aligned}
B_{out}(z = +L_c) &= \frac{A_{in} + B_{in}}{2} e^{-j\phi_{s,in}}e^{-j\phi_{s,out}}e^{-j2\beta_s L_c} \\
&\quad - \frac{A_{in} - B_{in}}{2} e^{-j\phi_{a,in}}e^{-j\phi_{a,out}}e^{-j2\beta_a L_c}.
\end{aligned}
\tag{6.9}
$$

Various distributions of A_{out} and B_{out} can be obtained by variations of L_c, A_{in} and B_{in}. In order to illustrate the power divider functions, we consider several examples.

(1) Whenever B_{in} is zero, the length of the two-mode waveguide L_c can be adjusted so that $\Delta\phi_Y = (\phi_{s,in} + \phi_{s,out} + 2\beta_s L_c) - (\phi_{a,in} + \phi_{a,out} + 2\beta_a L_c)$ is zero or $2n\pi$ so that

$B_{\text{out}} = 0$. This represents the case where all power is retained in waveguide A. Vice versa, when $\Delta\phi_Y$ is $(2n+1)\pi$, $A_{\text{out}} = 0$. This represents the case where all power is transferred from A to B.

(2) When A_{in} is zero, $A_{\text{out}} = 0$ when $\Delta\phi_Y = 0$ or $2n\pi$, while $B_{\text{out}} = 0$ when $\Delta\phi_Y = (2n+1)\pi$.

(3) When there is only input into either waveguide A or waveguide B, various relative distributions of power in the two output waveguides can be obtained by using an appropriate length of the two-mode waveguide to yield the necessary $\Delta\phi_Y$.[5]

(4) When there are inputs into both waveguides, the outputs depend on the relative magnitude of A_{in} and B_{in} and their relative phase. For example for $A_{\text{in}} = B_{\text{in}}$, only the symmetric mode is excited and A_{out} and B_{out} will always have equal magnitude.

6.1.1.5 The mode interference power divider

As a third example, the multimode interference coupler discussed in Section 2.4 is also a power splitter. It is illustrated in Fig. 2.10(a) for a two-mode interference coupler which has two input waveguides and two output waveguides. In this example, when only one of the waveguides is excited by the input radiation, the interference pattern of the total field in the coupler as a function of the distance was illustrated in Fig. 2.11.[6] If losses are neglected, the power transferred to any one of the output waveguides can be varied from 0% to 100% depending on the length of the coupler. A 2×2 InGaAsP MMI cross coupler has been made with $W = 8\,\mu\text{m}$ and $L = 500\,\mu\text{m}$ which gives excess loss of 0.4 to 0.7 dB and an extinction ratio of 28 dB. A 3 dB splitter has also been obtained with $L = 250\,\mu\text{m}$. The imbalance between the power in the two output waveguides is well below 0.1 dB. For a general multimode interference coupler, there are more than two modes in the coupler and more than two waveguides connected to the coupler. *When a waveguide is coupled to the input at an appropriate transverse position and output waveguides are coupled to the device after a specific propagation distance at their specific transverse positions, various distributions of power into the output guides can be obtained* [see 2].

Comparing all the power splitters, the Y-branch equal-power splitter is reliable and easy to make. The directional coupler is reliable and commonly used for unequal splitting of power into two outputs. It is difficult to control precisely the length of the two-mode waveguide in the back-to-back Y-branch coupler.[7] However, the relative phase of the guided-wave modes at the output is controlled. The Star coupler and the mode interference coupler provide $1 \times N$ splitting of power. The mode interference coupler is a very compact device. Typically its length is of the order of 100 μm. However, the Star coupler and the multimode interference coupler may have a higher insertion loss in comparison with the directional coupler or Y-branch.

6.1.2 Wavelength filters/multiplexers

In many applications such as wavelength division multiplexing, there are optical carrier radiations at slightly different wavelengths. Radiation within specific wavelength bands may need to be directed to different locations.

6.1.2.1   The reflection filter

In Section 2.2.3, the reflection of a forward propagating wave into the reverse direction by a grating fabricated in or on the waveguide was discussed. The reflectivity is wavelength sensitive. We showed in Eq. (2.18) that the 1st order reflection of a grating parallel to the waveguide is centered at the free space wavelength, $\lambda = 2n_{\text{eff}}T$, where T is the grating periodicity.[8] The wavelength bandwidth of the reflection is $\Delta\lambda = C_g n_{\text{eff}} T^2 / \pi$ on both sides of the center wavelength, where C_g is the coupling coefficient between the fundamental term of the Fourier series of the $\Delta\varepsilon$ of the grating and the electric field of the guided wave as shown in Eq. (2.15). The maximum magnitude of the reflection at the center wavelength is determined by the product of the coupling coefficient C_g and the length of the grating L. Occasionally, a higher nth diffraction order of the grating is utilized. Then its center wavelength will be at $\lambda = 2n_{\text{eff}}T/n$, and the C_g will be determined by the nth order Fourier coefficient of the $\Delta\varepsilon$ of the grating. A reflection filter has also been made in optical fibers in which the periodic $\Delta\varepsilon$ is fabricated in the cladding using the photo refractive effect. Note that if a grating is fabricated on or near a multimode waveguide, then the reflection property for each mode must be determined separately.

Wavelength filtering using a reflection grating in the planar waveguides was also discussed in Section 5.3.1. There is a great deal of similarity between the characteristics of the grating filter for the planar waveguide and for the channel waveguide. In comparison, all the transmitted and reflected signals are propagating in the same waveguide in the channel wavelength filters discussed in Section 2.2.3 and in the above paragraph. If the reflected radiation needs to be sent to another waveguide or fiber, an optical circulator will be used. However, in planar waveguides, the radiation can be incident on the grating at a small oblique angle of incidence, then the radiation within a specific wavelength range will be reflected to a slightly different direction. There is no need to use a circulator to separate the reflected beam from the incident beam. Note that the grating deflector discussed in Section 5.2.1.2 cannot be used as a wavelength filter because the angle and the magnitude of deflection are not sensitive enough to wavelength variation.

6.1.2.2   The phased array channel waveguide demultiplexer (PHASAR)

In order to direct optical radiations at different wavelength from an input waveguide into different output waveguides, let us consider a component called a PHASAR demultiplexer used in Wavelength Division Multiplexed (WDM) optical fiber systems [3]. Consider two Star couplers as discussed in Section 5.4 interconnected by an array of identical channel waveguides, each with length L_j, as shown in Fig. 6.1(a). On the input side of the first Star coupler, there is an input transmitting waveguide. The field distribution at the input is given by $E_y = \psi_k$ of the kth input channel and 0 elsewhere. In terms of the Star coupler discussed in Section 5.4, the input of the Star coupler consists of a circular array of channel waveguides on the input side within which only the kth waveguide (i.e. the transmitting waveguide) is excited. All other waveguides have zero power. The electric field of the input transmitting channel waveguide at the kth position will create a field distribution $E_y(R\theta')$ at the output circle in the first Star coupler. If all the interconnecting waveguides have equal length, and if the stray fields ζ in the gap between

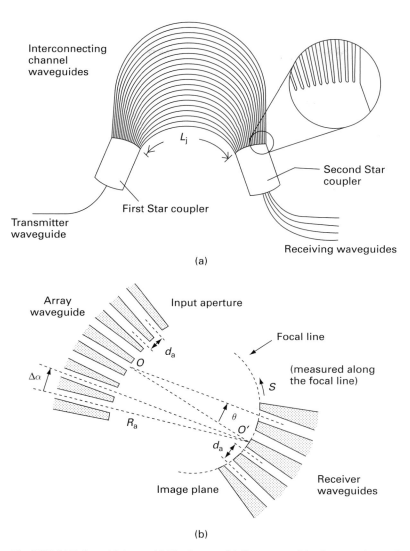

(a)

(b)

The PHASAR demultiplexer. (a) The layout. (b) Geometry of the Star coupler on the receiver side. (This copyright figure is taken from [3] with permission from IEEE.) The two Star couplers are connected by an array of interconnecting channel waveguides that have different lengths. Optical radiation from the input waveguide is transmitted to the interconnecting waveguides by the input Star coupler. The input radiation to the output Star coupler will have phase shifts controlled by the wavelength as well as by the length increments of the interconnecting waveguides. The objective is to create an appropriate phase shift so that radiation at different wavelengths is transmitted to different output waveguides of the receiver Star coupler.

channel guides are small, a field distribution identical to $E_y(R\theta)$ in the first coupler will be created on the input side of the second Star coupler. By reciprocity, this field distribution on the input side of the second Star coupler will create a field distribution on the output side which is the ψ_k at the position of the kth output waveguide and zero elsewhere. In other words, the power in the transmitting waveguide of the first coupler will now be transmitted exclusively to the kth output channel of the second Star coupler. The situation does not change if the length of the interconnecting waveguides between the two Star couplers differs so that the phase shift between adjacent interconnecting waveguides is 2π, i.e.

$$\frac{2\pi n_{\text{eff,c}}}{\lambda}\left(L_j - L_{j-1}\right) = \frac{2\pi n_{\text{eff,c}}}{\lambda}\Delta L = 2\pi. \tag{6.10}$$

Here, $n_{\text{eff,c}}$ is the effective index of the channel waveguide. The physical ΔL required to meet this condition will depend on λ.

Let the spacing between adjacent channel waveguides be d_a ($d_a = R\Delta\alpha$) in the first Star coupler. Then, according to Eq. (5.24), the E_y (created by the field of the kth channel waveguide in the input array) at the center of the mth waveguide in the output circular array, has a phase

$$\exp\left(\text{j}2\pi\frac{n_{\text{eff}}R}{\lambda}(k\Delta\alpha)(m\Delta\alpha)\right),$$

where $kR\Delta\alpha$ and $mR\Delta\alpha$ are the center angular positions of the kth and mth channel waveguide in the input and output array of the Star coupler as shown in Fig. 5.10 and Fig. 6.1(b). Note that k and m are integers, ranging from $-(N-1)/2$ to $(N-1)/2$, and n_{eff} is the effective index of the planar waveguide in the Star coupler. If the excitation changes from the kth waveguide to the $(k+1)$th waveguide, the difference of the E_y caused by this change is just a phase difference, $m\Delta\phi = 2\pi(R\Delta\alpha)(n_{\text{eff}}/\lambda)(m\Delta\alpha)$, at the center of the mth waveguide. Vice versa, when the radiation in the array of input waveguides in the second Star coupler has a total E_y field that contains this extra phase factor $m\Delta\phi$ for each input waveguide, $m = -(N-1)/2$ to $(N-1)/2$, the total radiation will be coupled to the $(k+1)$th output waveguide instead of the kth output waveguide.

The central idea of this demultiplexer is that, when the kth waveguide is the output guide at λ_1 and when the appropriate phase shift $m\Delta\phi$ is obtained as the wavelength is shifted from λ_1 to λ_2, we would have shifted the output from the kth waveguide to the $(k+1)$th waveguide at λ_2.

Let the difference in length of the adjacent interconnecting waveguides be ΔL. The mth interconnecting waveguide has a length $m\Delta L$ longer than the waveguide at the origin. Now consider in detail the second Star coupler at two different wavelengths, λ_1 and λ_2. Let the output channel be the kth waveguide at λ_1. This extra phase factor $m\Delta\phi$ (which is needed to shift the output to the $(k+1)$th waveguide) will be obtained at λ_2 when

$$m\Delta\phi = \frac{2\pi}{c}n_{\text{eff,c}}(\Delta f)m\Delta L, \quad \text{or} \quad \frac{R\Delta\alpha}{\Delta f} = \frac{d_a}{\Delta f} = \left(\frac{n_{\text{eff,c}}}{n_{\text{eff}}}\right)\left(\frac{\Delta L}{\Delta\alpha}\right)\frac{1}{f_2}. \tag{6.11}$$

Here, $f_1 = c/\lambda_1$, $f_2 = c/\lambda_2$, and $\Delta f = f_1 - f_2$. The ratio of $d_a/\Delta f$ is called the dispersion of the interconnecting waveguides. In practice, there may be optical carriers at a number of closely and equally spaced wavelengths, λ_1, λ_2, λ_3 ... (i.e. $\Delta f =$ constant), in the transmitting channel. When the above dispersion relationship is satisfied, optical carriers at different wavelengths are transmitted to a different output waveguide. This device is called a PHASAR wavelength demultiplexer in WDM fiber systems. The properties of the channel waveguides important to this application are the $n_{\text{eff,c}}$, the uniformity of $n_{\text{eff,c}}$ in different channels and the attenuation of the waveguides.

6.1.2.3   Resonator wavelength filters

Another class of narrow band wavelength filter is that of waveguide resonators. In order to discuss these device more thoroughly, they are presented in Section 6.1.4.

If we compare wavelength filters, it is clear that grating reflection filters and resonance filters are narrow band filters with well separated center frequencies. They have very similar properties which will be discussed again in Section 6.1.4. The PHASAR does not provide very narrow band wavelength filtering. It is primarily a wavelength multiplexer that direct signals at different equally spaced wavelengths into different channels.

6.1.3   Waveguide reflectors

If a channel waveguide is terminated abruptly at the free space end, the dielectric discontinuity will cause a reflection of the wave propagating in the waveguide. For a high index waveguide, this reflection could be quite large. This was used for reflection in semiconductor lasers in earlier days. In order to increase the reflectivity, a metallic film can be coated on a vertical cleaved end of a channel waveguide. The metal mirror will reflect all the incident radiation of the mode back into the backward direction of propagation independent of wavelength and mode pattern. However, the reflectivity of the metal is still not high enough for many applications, and it decreases slowly with shorter wavelength. Higher reflectivity mirrors can be made from multilayer dielectric coatings replacing the metal. The interference effect in the multilayer thin film creates a very high reflectivity that is wavelength selective.

In order to obtain end reflection, it is clear that vertical ends can only be made easily in some waveguides. In waveguides using a substrate such as GaAs, InP or Si, a vertical surface can be obtained by cleaving. In waveguides made from hard materials such as $LiNbO_3$, optically flat vertical ends can be obtained by mechanical polishing. Alternatively, the grating filter discussed in Sections 5.3.1 and 6.1.2.2 can be used as a reflector within the reflection bandwidth. It is called a distributed Bragg reflector (DBR). Grating reflectors are made on waveguide structures by etching or by optical exposure of the photo refractive material. Sometimes materials are deposited on the waveguide to fabricate the grating pattern. For a given λ, the larger the n_{eff} of the waveguide, the smaller is the required grating periodicity, and the harder it is to make the grating. The required grating periodicity is much longer if a higher order diffraction is used. In all these reflectors the guided wave is reflected back into the same mode in the same waveguide.

An unusual reflector would be the Y-branch reflector illustrated in Fig. 2.8(b). In this device, two identical single mode waveguides form the input of the Y-branch which feeds into a two-mode waveguide that has a specific length and a reflector at the end. It is discussed in Section 2.3.4.2(b). The reflector at the end can be any one of the reflectors discussed in the preceding paragraph. When there is radiation incident on one of the single mode waveguides in the input both the symmetric and the anti-symmetric mode are excited equally in the Y-branch and in the two-mode waveguide. The relative phases of the symmetric and anti-symmetric modes at the input are also equal so that the total field is concentrated only in the input waveguide before the Y intersection region. Since the symmetric and the anti-symmetric modes have the same phase velocity for isolated waveguides, this amplitude and phase relation is maintained until the two single mode waveguides become coupled (i.e. the symmetric and anti-symmetric modes have significant electric field between the two waveguides). In the coupled region and in the two-mode waveguide, the n_{eff} of the anti-symmetric mode is lower than that of the symmetric mode. The symmetric and anti-symmetric modes propagate adiabatically down the Y-branch, and they are reflected by the reflector and propagate back toward the input end. The total electric field pattern of the reflected modes at the input end is now determined by the relative phase of the two modes. Dependent on the relative phase, the total reflected field can be concentrated in the incident branch or in the other branch, or in some other ratio.[9] For a given fabricated Y-branch and end reflector, the relative phase in the transition region is fixed. Therefore the total relative phase (i.e. the ratio of the power reflected back into the single mode waveguides) is controlled only by the propagation length in the two-mode section. In a practical case, in a LiNbO₃ waveguide, the specific length of the two-mode waveguide section for a given desired reflection is obtained by polishing the two-mode waveguide to a specific length so that all the power is reflected back into the other waveguide with 2 dB of insertion loss [4].

In comparison, reflection by mirrors on the end of the waveguide is the simplest reflector to fabricate. Wavelength selectivity can be obtained with multilayer dielectric coating. However, for reflection within just a very narrow band of wavelength, a grating reflector such as DBR or DFR is required. A grating reflector is also used whenever a vertical end surface of the waveguide cannot be fabricated. The two-mode Y-branch reflector combines reflection with power division.

6.1.4   Resonators

Resonators have many important applications including intensifying the absorption or amplification effects, wavelength filtering and providing time delay.

6.1.4.1   The Fabry–Perot resonator

A Fabry–Perot optical resonator consists of a section of waveguide L long with a reflector at each end. There are multiple transmitted and reflected guided waves in the resonator. Resonance occurs when the round trip phase shift of the guided-wave mode between the two reflectors is multiples of 2π. Let R be the intensity reflectivity of the end reflectors, n_{eff} be the effective index of the mode propagating in the waveguide and α be the attenuation coefficient of the mode. Let us assume here that the amplitude reflection and transmission coefficients, r and t, at the input end and r' and t' at the output end are identical, with $R = rr' = rr = r'r'$. Note that r, r', t and t' are not wavelength dependent. The end reflectors are partially transmitting, with $R + tt' = 1$. Then the total transmission T of the power of the optical guided wave (i.e. P_{out}/P_{in}) through the resonator for a CW radiation at a single wavelength is [5]

$$T = \frac{P_{\text{out}}}{P_{\text{in}}} = \frac{(1-R)^2}{\left(1 - Re^{-2\alpha L}\right)^2 + 4Re^{-2\alpha L}\sin^2\delta}, \tag{6.12}$$

$$\delta = \frac{2\pi n_{\text{eff}}L}{\lambda} = \frac{\omega n_{\text{eff}}L}{c}, \tag{6.13}$$

where λ is the free space wavelength. Clearly T_{max}s occur at the wavelengths whenever $\delta = n\pi$, and T_{min}s occur at wavelengths where $\delta = (2n+1)\pi/2$. Since the sine function is non-linear, T in Eq. (6.12) for small α and large L is large only within a small range of λ around each T_{max}. These transmission peaks are known as the peaks of the Fabry–Perot resonance. The performance of the resonator is measured by: (1) its resonance frequencies; and (2) the frequency deviation from the resonance frequency (i.e. the linewidth) where T drops sharply.

The difference in frequency between two adjacent resonance peaks, called the Free Spectra Range (FSR), is

$$\text{FSR} = \omega_{n+1} - \omega_n = \frac{\pi c}{n_{\text{eff}}L}. \tag{6.14}$$

For $\alpha \cong 0$ and $R \cong 1$, if we let $\Delta\delta = \delta - n\pi$, then T at wavelengths near T_{max} can be expressed as

$$T = \frac{(1-R)^2}{(1-R)^2 + 4R\Delta\delta^2} \cong 1 - \frac{4R\Delta\delta^2}{(1-R)^2}. \tag{6.15}$$

Note that T drops to ½ when

$$(\delta - n\pi)^2 \cong \frac{(1-R)^2}{8R}. \tag{6.16}$$

If we let ω_{o} be the center of the resonance frequency, $\omega_{\text{o}} = n\pi c/n_{\text{eff}}L$, then the half linewidth $\Delta\omega$ in which the T drops to ½ is

$$\Delta\omega = \frac{(1-R)c}{2\sqrt{2R}n_{\text{eff}}L}. \tag{6.17}$$

A measure commonly used to gauge the resonance linewidth in the optics literature is the Finesse, F, which is FSR divided by the full linewidth. Using the half linewidth $\Delta\omega$ in Eq. (6.17), we obtain

$$F = \frac{\sqrt{2R}\pi}{(1-R)c}. \tag{6.18}$$

Resonators are also rated in terms of their Q factor in the electrical engineering literature. The Q factor calculated for the Fabry–Perot resonance with $\alpha \cong 0$ and $R \cong 1$ is

$$Q = \varpi_{\text{resonance}}\frac{\text{field energy stored}}{\text{power dissipated}} = \frac{n\pi}{1-R}. \tag{6.19}$$

Note that Q is also a measure of resonance full linewidth $\Delta\omega$, $1/Q = \Delta\omega_{\text{full}}/\omega_{\text{o}}$. From Q, we obtain[10]

$$\Delta\omega_{\text{full}} = \frac{(1-R)c}{n_{\text{eff}}L}. \tag{6.20}$$

A Fabry–Perot resonator is most commonly used as a narrow band wavelength filter centered about its resonance frequencies. There are many T_{max}s at different resonance frequencies. When peak transmission at only a specific wavelength is desired, additional wavelength filtering of R and T could be created by using a multilayer dielectric coating or a grating reflector, which was discussed in Section 2.2.3. For example, in lasers, amplification takes the place of the attenuation α, and laser oscillations will occur at all the Fabry–Perot resonance wavelengths within the amplification linewidth of the material. These are known as the longitudinal modes of the laser oscillation. In single mode lasers, additional frequency selectivity is provided by distributed Bragg reflectors or distributed feedback gratings to eliminate unwanted longitudinal modes of oscillation. When the Fabry–Perot resonators are compared with the resonances in grating filters, the wavelength selectivity of the Fabry–Perot resonance with small α, large R and large L/λ, can be much higher than that of a grating.

In addition to resonance filtering, the Fabry–Perot resonance effect is useful also in other applications. When T is measured as a function of the wavelength for a Fabry–Perot resonator with a known length L, then the $\Delta\lambda$ of the adjacent resonances gives an accurate determination of n_{eff}. Let λ_1 and λ_2 be the wavelength of two adjacent peaks, then the n_{eff} of the waveguide is related to $\Delta\lambda$ by

$$2n_{\text{eff}}L = \lambda_2\lambda_2/|\lambda_1 - \lambda_2|. \tag{6.21}$$

On the other hand, the measurement of the ratio of T_{max} to T_{min}, known as the contrast ratio CR ($CR = T_{\text{max}}/T_{\text{min}}$), can be used to determine the α,

$$\alpha = \frac{1}{2\text{L}}\ln\left[\frac{1}{R}\frac{\left(\sqrt{CR}+1\right)}{\left(\sqrt{CR}-1\right)}\right]. \tag{6.22}$$

In general, any two discontinuities on a waveguide could cause significant reflections between them. For example a section of cleaved waveguide will have reflections at the ends due to dielectric discontinuity. The reflections will create a mild Fabry–Perot resonance effect which may implicitly affect the device performance significantly.

6.1.4.2   The ring resonator

Channel waveguides could also be made into a ring (or loop) as illustrated in Fig. 6.2(a). Resonances in an isolated waveguide ring occur at frequencies ω when the phase shift of a guided-wave mode after one round of propagation is a multiple of 2π,

$$\frac{\omega n_{\text{eff}}}{c}(2\pi R) = 2n\pi, \tag{6.23}$$

where R is the radius of the ring. However, the ring resonance is useless unless a guided wave in another waveguide is coupled to the guided wave in the ring. A ring resonator

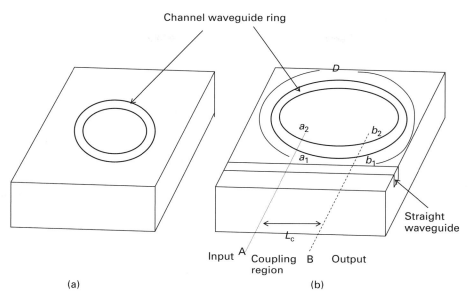

Channel waveguide ring

D

a_2 b_2

a_1 b_1

Straight waveguide

L_c

Input A Coupling B Output
region

(a)   (b)

Fig. 6.2. Ring resonators. (a) An isolated ring resonator. (b) A ring resonator coupled to a straight input-output waveguide.

coupled to a straight waveguide via a variable gap directional coupling interaction is illustrated in Fig. 6.2(b).

Directional coupling between two adjacent waveguides was discussed in Section 2.2.4 for a constant coupling gap. Directional coupling between two waveguides with a variable coupling gap can be approximated as a cascade of short directional couplers which has a constant local coupling gap within each local section. The total coupling effects can be calculated from all the short local sections as follows. Results obtained in Eq. (2.22) and (2.23) for two coupled waveguides could be rewritten in matrix form for the jth local section with constant coupling gap as

$$\begin{vmatrix} b_{1j} \\ b_{2j} \end{vmatrix} = \begin{vmatrix} t_j & \kappa_j \\ \kappa_j^* & t_j^* \end{vmatrix} \begin{vmatrix} a_{1j} \\ a_{2j} \end{vmatrix}, \tag{6.24}$$

where a_{1j} and b_{1j} are the complex amplitude of the guided-wave at the input and output of the straight waveguide, while a_{2j} and b_{2j} are for the ring waveguide. At the junction between the jth local section and the $(j+1)$th local section, $a_{1,j+1} = b_{1j}$ and $a_{2,j+1} = b_{2j}$. Therefore, the total effect of the variable coupling can be expressed as a matrix $\begin{vmatrix} t & k \\ k^* & t^* \end{vmatrix}$ which is the product of all these $[t_j \ \kappa_j]$ matrices. Alternatively the coupled mode equation can be solved with a variable coupling coefficient C.

In Fig. 6.2(b), A marks the beginning of the coupling region and B marks the end of the coupling region. The distance between A and B of the coupling region is L_c. The length of the isolated waveguide in the ring is D. The incident optical guided wave in the straight waveguide is shown to have complex amplitude a_1 at position A before the coupling. The exit optical wave in the straight waveguide is shown to have complex amplitude b_1 at position B. The complex amplitude of the guided wave in the ring resonator is a_2 at A and b_2 at B. Note that a_1, a_2, b_1 and b_2 are related by the variable coupler as

$$\begin{vmatrix} b_1 \\ b_2 \end{vmatrix} = \begin{vmatrix} t & \kappa \\ \kappa^* & t^* \end{vmatrix} \begin{vmatrix} a_1 \\ a_2 \end{vmatrix},$$ (6.25)

where

$$\kappa\kappa^* + tt^* = 1.$$ (6.26)

On the other hand, for the guided-wave propagating from B to A in the ring, the distance of propagation is D in the isolated waveguide. Therefore

$$a_2 = b_2 e^{-\alpha D} e^{-jn_{\text{eff}}kD} = b_2 e^{-\alpha D} e^{-j\theta},$$ (6.27)

where α is the attenuation coefficient of the guided-wave mode and n_{eff} is its effective index. Note that, from Eq. (6.25), $b_2 = \kappa^* a_1 + t^* a_2$ where $t = |t|e^{j\phi_t}$. Thus the phase shift for one round of propagation is $\theta + \phi_t$. Similarly to what is shown in Eq. (6.23), resonance for a CW radiation at a single free space wavelength λ_o now occurs when $\theta + \phi_t = 2n\pi$ which is[11]

$$\theta_o + \phi_t = \frac{2\pi n_{\text{eff}}(D + L_c)}{\lambda_o} = 2n\pi.$$ (6.28)

The power transmitted from the input guided wave a_1 to the output is $|b_1/a_1|^2$, and the power transmitted from the input guided wave a_1 to the re-circulating guided wave in the ring is $|a_2/a_1|^2$. They have been calculated from Eq. (6.25) by Yariv [6] to be

$$\left|\frac{b_1}{a_1}\right|^2 = \frac{e^{-2\alpha D} + |t|^2 - 2e^{-\alpha D}|t|\cos(\theta + \phi_t)}{1 + e^{-2\alpha D}|t|^2 - 2e^{-\alpha D}|t|\cos(\theta + \phi_t)},$$ (6.29)

$$\left|\frac{a_2}{a_1}\right|^2 = \frac{e^{-2\alpha D}\left(1 - |t|^2\right)}{1 - 2e^{-\alpha D}|t|\cos(\theta + \phi_t) + e^{-2\alpha D}|t|^2}.$$ (6.30)

At resonance, $\cos(\theta + \phi_t) = 1$ while $\left|\frac{b_1}{a_1}\right|^2$ drops to zero when $e^{-\alpha D} = |t|$, known as the critical coupling condition. At critical coupling (i.e. $\exp(-\alpha D) = |t|$), there is perfect destructive interference between the guided-wave in the output waveguide coupled from the ring and from the input. The amplitude $\left|\frac{a_2}{a_1}\right|^2$ also soars to a high value near resonance. Its maximum value at resonance is

$$\left|\frac{a_2}{a_1}\right|^2_{\max} = \frac{|t|^2}{1 - |t|^2}.$$ (6.31)

The FSR of adjacent resonances at wavelengths, λ_n and λ_{n+1}, is

$$\text{FSR} = \omega_{n+1} - \omega_n = \frac{2c\pi}{n_{\text{eff}}(D + L_c)}. \tag{6.32}$$

As the wavelength changes, "$\theta + \phi_t$" deviates from its resonance condition by

$$\Delta(\theta + \phi_t) \cong \frac{n_{\text{eff}}(D + L_c)(\omega - \omega_o)}{c}. \tag{6.33}$$

As $\Delta(\theta + \phi_t)$ increases, $\left|\dfrac{b_1}{a_1}\right|$ will increase and $\left|\dfrac{a_2}{a_1}\right|$ will decrease. When there is critical coupling, $\left|\dfrac{a_2}{a_1}\right|^2$ drops to half of its maximum value when

$$\Delta\omega = \frac{\left(1 - |t|^2\right)c}{n_{\text{eff}}D|t|} = \frac{|\kappa|^2 c}{n_{\text{eff}}D|t|}. \tag{6.34}$$

Here $\Delta\omega$ is defined as half linewidth when the intensity drops to ½ of its maximum. Again, we can calculate the Q factor of the resonator to be

$$Q = \frac{\omega_o}{2\Delta\omega} = \frac{\omega_o n_{\text{eff}}D|t|}{2c|\kappa|^2}. \tag{6.35}$$

Assuming that $L_c \ll D$, the finesse of the ring resonator is approximately

$$F = \frac{\omega_{n+1} - \omega_n}{2\Delta\omega} \cong \frac{\pi|t|}{\left(1 - |t|^2\right)}. \tag{6.36}$$

There are many similarities between the ring resonator and the Fabry–Perot resonator. For comparison, let us assume that the R in the Fabry–Perot resonator is independent of frequency and that the coupling κ in the ring resonator is also independent of frequency. Both of them will have sharp resonances equally spaced in frequency. It is also interesting to compare the differences. The important characteristics to be compared are the wavelength selectivity (i.e. the FSR) and the linewidth or the Q of the resonances. (1) It is hard to get very long low loss waveguides for Fabry–Perot resonators. It means that D of the ring resonator tends to be much bigger than the L of the Fabry–Perot resonator. The reflectivity of mirrors in Fabry–Perot resonators is limited. It is easy to get very high |t| and low propagation loss in ring resonators. Thus it is easy to achieve very high Q or small linewidth in ring resonators at critical coupling. (2) On the other hand, the radiation loss of channel waveguide ring resonators increases with decrease of the radius of curvature. Therefore, the FSR, i.e. the separation of the adjacent resonances, is usually much larger in Fabry–Perot resonators than in ring resonators. Although high Q (or F) is very important in many filtering applications, FSR also needs to be large in many other applications. Kominato et al. have shown that a finesse F larger than 30 has been obtained in ring resonators made from GeO₂ doped silica waveguides with a ring radius of 6.5 mm at λ = 1.55 μm [7]. However, the FSR of their resonator is only 5 GHz.

(3) In Fabry–Perot resonators, additional wavelength selectivity can be obtained by using a grating filter or multilayer dielectric coating for reflection (or feedback) so that the Q of the adjacent resonances is wavelength selective. It is not possible to have a high wavelength selectivity in directional coupling (i.e. κ). Therefore a technique such as a double ring resonator is used to achieve wide FSR [8]. A double ring resonator with 100 GHz of FSR and F > 138 was demonstrated by Suzuki et al. [9].

High Q or high Finesse depends on the ability to achieve small α in the ring. There are usually two major factors that may contribute to the propagation loss: (1) the scattering and absorption loss of the channel waveguide; and (2) the radiation loss of a curved waveguide. When there is a bend in the waveguide there will be radiation loss into the cladding or substrate.

The radiation loss of a curved waveguide

In a straight planar waveguide, the guided-wave mode is considered as plane waves totally internal reflected at the boundaries of the core layer. In a straight channel waveguide, the guided-wave mode can be considered as planar guided waves totally reflected at the lateral boundaries of the core. There is an evanescent field in all the cladding regions because the β_z (i.e. $n_{eff}kz$) in the direction of propagation is so large that the propagation wave numbers of the fields in the lateral directions in the cladding are imaginary, as they are required by the continuity of the fields in the longitudinal direction. Total internal reflection has zero propagation loss in the cladding regions as long as the propagation wave numbers in the lateral directions, β_x and β_y, are imaginary. When the waveguide is curved with a curvature ρ, the lateral region outside the curved waveguide fans out. The electromagnetic field in the expanded lateral region propagates with a new expanded coordinate in the z direction. The expansion of the coordinate in the z direction increases as the distance from the waveguide increases. At some distance away from the waveguide, the β of the fields in the lateral direction outside the curve no longer needs to be imaginary in order to meet the continuity condition of $n_{eff}kz$. At this point, the fields become propagating waves. Energy will radiate away. The total internal reflection will now have a radiation loss. The smaller the curvature ρ, the larger the radiation loss. Unger has presented clearly an analysis of the radiation loss in a curved planar waveguide [10]. His analysis shows that the radiation loss increases exponentially as $k\rho$ is decreased. Kominato *et al.* [7] have shown experimentally that the radiation loss increased dramatically in their waveguides for bending radius less than 4 mm.

The propagation loss

There are two kinds of propagation loss in waveguides, absorption loss and scattering loss. Volume scattering is usually caused by defects in materials while surface scattering is caused by defects on the interface created during processing. Low loss straight channel waveguides have been made in LiNbO$_3$ waveguides by diffusion, without surface scattering as discussed in Section 1.3.1. However, the propagation loss of curved LiNbO$_3$ waveguides is generally unknown. Absorption loss occurs in semiconductors due to dopands and free carriers. Although absorption in intrinsic semiconductors can be kept very low, substantial surface scattering loss occurs quite often in channel wave-guides in high refractive index crystalline media because of the defects produced in fabrication processes. For this reason, low loss semiconductor waveguides are usually ridged waveguides as discussed in Section 1.3.2. Surface scattering loss is especially high in curved semiconductor waveguides because etching tends to follow crystalline orientation, thereby creating large defects along the curved boundary. Therefore low loss ring resonators are made primarily in doped silica waveguides on Si substrates, as discussed in Section 1.3.4.

6.1.5 The optical time delay line

Controlled true time delay of optical signals has many important applications. For example, in optical time division multiplexed communication this function is required for synchronization of optical signals or buffering of optical data. In a phased array antenna, variation of the delay of the RF signals to different antenna elements can control the radiation direction of the antenna. Since RF signals can be carried on the optical carrier, a time delay of RF signals is accomplished by a time delay of optical signals.

An optical signal group delay, τ_g, is obtained when the signal propagates in a waveguide or an optical fiber for a distance L with group velocity v_g, $\tau_g = L/v_g$. The difficulties of this approach are that a long τ_g requires long L and that a different τ_g requires a different length. For example, 1 μs of delay requires 200 m of fiber. Therefore, Fabry–Perot and ring resonators have been used to obtain group delays that are many times longer than L/v_g.

When a pulsed optical signal is injected into a resonator as discussed in Section 6.1.4, it is reflected back and forth in a Fabry–Perot resonator and re-circulated in a ring cavity. The optical signal pulse is transmitted periodically to the output whenever it reaches the output port. Therefore, there are delayed output optical signals at multiple delay time intervals of $2L/v_g$ in Fabry–Perot resonators (or $(D+L_c)/v_g$ in ring resonators). For low loss resonators, the output pulses will repeat many times. If there are n_g output pulses and if the last pulse is used for signal processing then the total available time delay of this pulse from the input pulse is n_g times the single time delay interval of the resonator. The time response of a resonator is related directly to the frequency response of the resonator (e.g. FSR) discussed in the previous section.

The relation between the time behavior and the frequency spectrum of any output field can be illustrated through a very simple example. It is well known that when there are N output fields at discrete frequencies separated at an equal frequency interval $\delta\omega$ around a center frequency ω_o, we obtain mathematically

$$E = \sum_{-(N-1)/2}^{+(N-1)/2} A e^{j(\omega_o + n\delta\omega + \phi)t} = A e^{j\omega_o t} e^{j\phi} \frac{\sin(N\delta\omega t/2)}{\sin(\delta\omega t/2)}, \qquad (6.37)$$

where A and ϕ are amplitudes and phase of all the output fields, n identifies the individual field at frequency $\omega_o + n\delta\omega$, n varies from $-(N-1)/2$ to $+(N-1)/2$ for odd N, and E is now periodic in t with period $T = 2\pi/\delta\omega$. This example assumes that there is no loss of amplitude from pulse to pulse and that the linewidths of the resonances are infinitely narrow. It is a useless example because outputs with zero bandwidth carry no information. Sufficient linewidth is required to accommodate inputs that have finite frequency spectra. Nevertheless, the example illustrates clearly the direct relation between frequency response and time response. In reality, because of the α in ring resonators (or the mirror transmittance or loss in Fabry–Perot resonators), the linewidth of resonances broadens and n_g decreases. The behavior of resonators for time delay is quite complicated. For different applications, there are trade-offs between time delay and pulse distortion that need to be made.

More realistically, a resonator is a linear time-invariant system. Therefore its output is characterized mathematically by its impulse response function $h(t)$. Given an input signal $x(t)$, its output signal $y(t)$ is

$$y(t) = \int x(\tau)h(t - \tau)\mathrm{d}\tau. \tag{6.38}$$

Equivalently, in the frequency domain, we have

$$Y(\omega) = H(\omega)X(\omega), \tag{6.39}$$

where $X(\omega)$, $Y(\omega)$ and $H(\omega)$ are related to $x(t)$, $y(t)$ and $h(t)$ by a Fourier transform. The trade-offs between frequency and time responses are governed by these transform relations. Lens *et al.* have presented a detailed review of the analysis of the trade-offs between bandwidth constraint and time delay in optical resonators [11]. In their analysis they have also shown that the trade-offs are different for Fabry–Perot resonators and for ring resonators. The limitation of bandwidth and time delay in a single resonator can be circumvented by using a cascade of resonators tuned to slightly different resonance frequencies to broaden the total bandwidth. Zhuang *et al.* have demonstrated a maximum delay of 1.2 ns over a bandwidth of 2 GHz with a group delay ripple below 0.1 ns, using a cascade of seven ring resonators. Using different combinations of various numbers of ring resonators, they also have obtained time delays of 0.1, 0.25, 0.5, 0.7, 0.8 and 1.0 ns [12].

6.2   Active waveguide components

Besides lasers and detectors, modulators and switches are the most common active waveguide devices. Although modulation and switching can also be accomplished using acousto-optical effects,[12] magnetostrictive effects, electro-mechanical effects and even thermal effects, high speed and high frequency switching and modulation have been achieved only by electro-optical effects such as discussed in Chapter 3. The electrical characteristics of electro-optical modulators and switches are as important as their optical characteristics in all applications. The design and analysis of these devices differ significantly for high speed (or high frequency) operation and for moderate and low speed (or frequency) operation. *For low and moderate speed (or frequency) operations or short devices, the electrodes producing the electro-optical effects are shorter than the wavelength of the electrical signals controlling them. The transit time for the optical signal to propagate through the device is much shorter than the time period of the electrical control signal. Thus the instantaneous electric field at different locations of the device seen by the optical guided wave has approximately a uniform time variation. In Section 4.1.1 we showed that, in this case, these devices should be represented by lumped element circuit elements in the analysis of the driving circuit. For high speed (or high frequency) operations or long devices, the electrical current and voltage on the electrodes are now a function of both time and position. We showed in Section 4.2 that the electrical voltages and currents now travel in the electrodes as they are in an electrical transmission line. The analysis of the electro-optical modulation (or switching) characteristics must take into account the electro-optical interaction between a traveling optical wave and a traveling electric wave, as discussed in Section 4.3.*

For digital applications, what is important is the time varying characteristic of the modulators or switches in response to a large voltage pulse. As we have discussed in the introductory section of Chapter 4, the time response characteristic of the devices is often non-linear with respect to the amplitude of the applied electrical pulse. It is difficult to discuss the time response of a large voltage pulse in general. However, pulse response can often be inferred from the device response to small signals at various frequencies. Therefore discussion on the time response of active devices in this chapter will be focused on responses to small RF signals at various frequencies.

The presentation on active channel waveguide devices will be divided first into the electrical circuit behavior of lumped element devices that have a length shorter than the electrical wavelength of the highest operating frequency, and longer traveling wave devices that have a device length comparable to the electrical wavelength at the high operating frequencies. Within each type, lumped element or traveling wave device, devices that operate under different types of guided-wave interaction and electro-optical effect will then be discussed. Modulators and switches made with the same type of guided-wave and electric field interaction have many characteristics in common. Thus discussion on how a given type of modulator can function as a switch appears at the end of each subsection.

6.2.1   Lumped element modulators and switches

At low frequencies or within devices much shorter than the electrical wavelength, the time variation of electrical voltage and current at different positions on the electrode is identical. The transit time of the optical wave propagating through the device is very fast compared to the time variation of the electrical control signal. The instantaneous electric field at different longitudinal positions which creates the electro-optical effect experienced by the propagating optical wave has the same time variation. We have shown in Section 4.1.1 that these devices behave electrically as lumped elements, R_m, L_m and C_m, in the driver electric circuit. In most cases, ωL_m is negligible. The internal resistance R_{source} of the driving source is 50 Ω. The low frequency bandwidth spreads from DC to the upper frequency limit in which the device response falls to 50% of its maximum

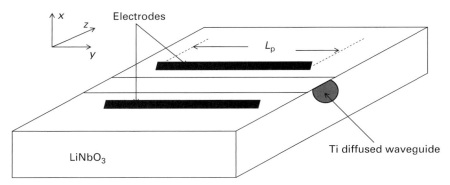

Fig. 6.3.   x-cut LiNbO$_3$ phase modulator.

response at DC. The maximum operating frequency is limited by the $R_{source}C_m$ constant of the driving circuit. The smaller the device capacitance, the larger is the bandwidth. When we compare different devices for a given application, the device capacitance C_m is the most important consideration that limits the operation bandwidth of the device. Impedance matching is needed to transfer a maximum amount of the driver RF power to the device.

6.2.1.1   The phase modulator

The phase modulator is a very simple device. Phase modulation of a guided wave is obtained when an electric field is applied to a waveguide fabricated on electro-optic materials. According to Section 2.1.1, the electric field of the mth guided-wave mode propagating in the z direction is described as

$$\boldsymbol{E}(\boldsymbol{x},\boldsymbol{y},\boldsymbol{z}) = A\boldsymbol{e_m}(\boldsymbol{x},\boldsymbol{y})\mathrm{e}^{-\frac{a_g}{2}z}\mathrm{e}^{-\mathrm{j}(n_{\mathrm{eff}}+\Delta n_{\mathrm{eff}})kz}, \tag{6.40}$$

where α_g is the power attenuation coefficient of the propagating mode, n_{eff} is the effective index of the mode in the absence of the applied electric field, and Δn_{eff} is the instantaneous change of the effective index produced by the applied electric field $\boldsymbol{F}$.[13] Figure 6.3 illustrates a phase modulator fabricated on x-cut LiNbO$_3$. Phase modulators have also been made in other waveguide materials such as polymers and III-V semiconductors. In all linear electro-optic materials, Δn_{eff} is directly proportional to the applied electric field $\boldsymbol{F}(\boldsymbol{x},\boldsymbol{y})$ which is proportional to the applied voltage V on the electrode. Thus Δn_{eff} is directly proportional to the instantaneous V. For a uniform electrode L_p long and in the lumped element approximation, the accumulated change in the phase of $E(x,y,z)$ produced by the electro-optic effect is

$$\Delta\phi = \Delta n_{\mathrm{eff}}(V)kL_p = \Delta n_{\mathrm{eff}}(V)|_{V=1}kVL_p. \tag{6.41}$$

In all these modulators, Δn_{eff} could be calculated from $\Delta n(x,y)$ by solving the guided-wave mode equation as a function of $\Delta n(x,y)$. Alternatively, it could be calculated from Δn using the perturbation method discussed in Sections 2.1 and 3.5.1. From the perturbation analysis, we obtain:

$$\Delta n_{\text{eff},m} = \frac{n}{n_{\text{eff},m}} \Gamma \Delta n_{\text{av}} = \frac{n}{n_{\text{eff},m}} \frac{\displaystyle\iint_{\text{electro-optic region}} \Delta n \underline{e_m} \bullet \underline{e_m^*} dx\, dy}{\displaystyle\int_{-\infty}^{\infty} \int_{-\infty}^{\infty} \underline{e_m} \bullet \underline{e_m^*} dx\, dy}. \tag{6.42}$$

Section 3.1 discussed how the Δn of the material is determined from the applied electric field $\underline{F}$. Some examples of $\Delta n(\underline{F})$ for LiNbO$_3$, the polymer and the III-V compound semiconductor materials are shown in Eq. (3.19) and (3.22).

Note that $\underline{F}$ is produced by the voltage applied to the electrodes. The electrodes are usually made in several common configurations such as micro-strip line, coplanar waveguide and coplanar strip. Figure 1.9 showed the electrode configuration in the form of a micro-strip transmission line for polymer waveguides. Figure 3.2(a) showed the electrode configuration for phase modulators fabricated on z-cut LiNbO$_3$. Although the concept of electro-optic phase modulation is simple, the calculation of $\underline{F}(x,y)$ from a given applied voltage V on the electrode could be complex. A crude common macroscopic approximation is to regard the electric field between two electrodes in terms of a constant F between two parallel plate electrodes separated by a gap distance d.[14] In that case

$$F \approx V/d. \tag{6.43}$$

Here d is the separation of the electrodes on insulating waveguides or the thickness of the intrinsic layers in p–i–n structures. Note that $\underline{F}$ is oriented in the direction from the electrode at lower potential to the electrode at higher potential. The parallel plate electrode approximation is so popular that the complex actual field $\underline{F}(x,y)$ of the electrodes is frequently written as

$$\underline{F}(x,y) = \frac{V}{d} \underline{f_F}(x,y), \tag{6.44}$$

where $\underline{f_F}$ is used to modify the parallel plate approximation to represent the actual $\underline{F}$. From Eq. (6.42), (6.43), and (6.44), it is clear that Δn_{eff} is directly proportional to V. The term $\Delta n_{\text{eff}}|_{V=1}$ in Eq. (6.41) is the proportionality constant.

Fortunately, the electromagnetic fields of any two-conductor electrodes are approximately TEM fields. Thus $\underline{F}(x,y)$ can be calculated using an electrostatic approximation and the Laplace equation. The calculation of such a transmission line on insulating materials has been discussed in many microwave publications [see 13 and 14]. However, in a phase modulator made from semiconductor optical waveguides, the electrode structure frequently contains a p–i–n diode. The electro-optic change of index is created in the thin intrinsic layer of the reverse biased p–i–n junction so that a large F could be created from a small applied voltage V. Figure 3.3 shows such an example. In this case the calculation of F in the intrinsic region involves calculation of fields and voltages across doped layers and heterojunctions. There are numerical simulation programs such as *ANSOFT* that will calculate $\underline{F}$ for a given electrode and waveguide configuration for various material and electrode structures, including the semiconductor heterojunctions.

The applied electric field is proportional to the voltage applied to the electrodes. When a RF source is used to drive the phase modulator, the electrodes can be best represented by a capacitance C_m in the driving circuit. In the parallel plate approximation, the capacitance C_m of the modulator using electrodes in the micro-strip configuration is

$$C_m \approx \frac{\varepsilon W}{d} L_\mathrm{p}, \tag{6.45}$$

where W is the width of the electrode, and $\varepsilon W/d$ is the capacitance per unit length of the electrode. There is no simple approximation for calculating the capacitance per unit length in coplanar waveguide or coplanar strip configurations. However, C_m can be calculated directly from $\underline{F}$. For example, one way to calculate C_m is to use Eq. (4.3).

The performance of any phase modulator is measured in several different ways. (1) A modulator is rated by the voltage required to achieve a given $\Delta\phi$ such as π. From Eq. (6.41) it is clear that the V required for a given optical $\Delta\phi$ is inversely proportional to L_p. The longer the L_p, the smaller is the required V. From Eq. (6.43), the smaller the d, the smaller is the required V. However, d will also affect $\Gamma\Delta n_\mathrm{av}$. (2) A modulator is rated by the bandwidth within which the modulator can be driven effectively by a RF source. According to Section 4.1, when such a capacitance is driven directly by a RF source the frequency bandwidth of the modulator is $1/R_\mathrm{source}C_m$. Note that C_m is directly proportional to L_p. Impedance matching is often used to improve the efficiency with which the RF power is transmitted to the modulator. For example, if impedance matching is provided by a 50 Ω resistor in parallel with C_m, then the low-pass bandwidth of the modulator is limited to $\omega = 2/R_\mathrm{source}C_m$. Different matching arrangements will modify the bandwidth differently. No matter what impedance matching method is used, the larger the L_p and the smaller the d, the larger is the C_m, and the smaller will be the bandwidth. There is a trade-off between the voltage required to obtain a required $\Delta\phi$ and the bandwidth. As an example, the advantage of using a thin intrinsic layer in a p–i–n junction to reduce the modulation voltage is balanced by the disadvantage of a large capacitance created by the small d.

Phase modulation is not used much because it is not easy to detect a change in optical phase. Most commonly, interference with another optical wave is used to convert the phase modulation into a modulation of the intensity of the interference pattern. An example is the Mach–Zehnder modulator. *The major advantage of phase modulators is that the $\Delta\phi$ is strictly linearly proportional to V. A stand-alone phase modulator cannot function as a switch, but phase modulation could be utilized in coordination with other devices such as the Y-branch power divider discussed in Section 6.1.1.4 to produce switching.*

6.2.1.2   The Mach–Zehnder modulator/switch

A typical Mach–Zehnder (MZ) interferometer was illustrated in Fig. 2.9. Its principle of operation was discussed in Section 2.3.4. It consists of single mode input and output waveguides connected to two symmetrical Y-branch couplers back-to-back. The Y-branch couplers are interconnected by two identical single mode waveguides. The

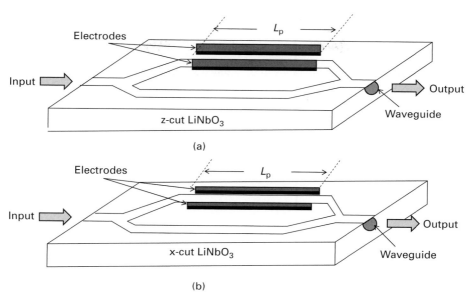

Fig. 6.4. A Mach-Zehnder modulator on LiNbO₃ substrate. (a) MZ modulator on z-cut LiNbO₃. (b) MZ modulator on x-cut LiNbO₃. Note that L_p is the length of the electrode producing the phase modulation. Phase modulation in one arm of the Mach-Zehnder modulator is illustrated here. When push-pull phase modulation is applied to both arms the V_π of the modulator will be reduced by 50%.

single mode connecting waveguides are well isolated from each other so that their evanescent fields no not overlap. When a MZ interferometer is fabricated on an electro-optic material, the n_{eff} of the guided-wave mode in the interconnecting waveguide can be modulated by the electro-optic effect. In essence, a MZ modulator is just two phase modulators connected to two Y-branch couplers.

Mach-Zehnder (MZ) modulators have been used in many applications, and such modulators in LiNbO₃ are commercially available. Figure 6.4(a) illustrates schematically a MZ modulator on z-cut LiNbO₃, while a MZ modulator on x-cut LiNbO₃ material is illustrated in Fig. 6.4(b).[15] In both cases, the input Y-branch splits the input guided wave into guided waves in two identical connecting waveguides, called the arms. In an ideal MZ modulator, these two guided waves have equal amplitude and identical phase in the absence of electro-optic modulation. When they propagate to the output Y-branch, they excite the symmetric mode which transforms adiabatically into the fundamental mode of the single mode output waveguide. Therefore, neglecting any losses, all the input power is transmitted to the output. When a modulation voltage is applied to the electrodes, the n_{eff} of the guided wave is changed by the electro-optic effect. Thus the arm with the electrode functions like a phase modulator. In order to take advantage of the large electro-optic coefficient, r_{33} in LiNbO₃, the electric field is applied in the z direction of the material. The parallel electrodes are L_p long. According to the perturbation analysis discussed in Section 3.5.1 and Eq. (3.38), (6.41) and (6.42), the $\Delta\phi$ in any arm created by the V is:

$$\Delta\phi = \pi V \frac{n_e^3 r_{33} L_p}{\lambda d} \left(\frac{n_e}{n_{eff}}\right) \left[\frac{\iint f_F(x,y)\underline{e}(x,y)\bullet\underline{e}^*(x,y)\mathrm{d}x\,\mathrm{d}y}{\iint \underline{e}(x,y)\bullet\underline{e}^*(x,y)\mathrm{d}x\,\mathrm{d}y}\right] = \frac{\pi V}{V_\pi}. \tag{6.46}$$

When $\Delta\phi = n\pi$, the fields of the guided waves in the two arms incident on the output Y-branch are $180°$ out of phase. They excite the anti-symmetric mode. The anti-symmetric mode is cut-off in the output single mode waveguide. All its power is transmitted to the radiation modes at the cut-off point of the anti-symmetric mode in the Y-junction. Thus no power is transmitted to the output. From Eq. (2.32), we obtain

$$I = \frac{I_o}{2}(1 + \cos\Delta\phi) = \frac{I_o}{2}\left(1 + \cos\left(\frac{\pi}{V_\pi}V\right)\right), \tag{6.47}$$

where I_o is the input power to the MZ modulator. From Eq. (6.46), we obtain:

$$V_\pi L_p = \frac{\lambda d n_{eff}}{n_e^4 r_{33}}\left[\frac{\iint f_F(x,y)\underline{e}\bullet\underline{e}^*\mathrm{d}x\,\mathrm{d}y}{\iint \underline{e}\bullet\underline{e}^*\mathrm{d}x\,\mathrm{d}y}\right]^{-1}. \tag{6.48}$$

In the push-pull version of MZ modulators, there are electrodes on both arms. The voltages applied to electrodes on two arms have opposite polarity, the electric field creates a $+\Delta n_{eff}$ in one arm and a negative Δn_{eff} in the other. In short, both arms are phase modulators with opposite $\Delta\phi$. Therefore, the required $\Delta\phi$ in each arm for zero output power is only $n\pi/2$, and the V_π is ½ of the V_π in Eq. (6.48) derived for modulators using only one arm.[16] Equation (6.47) still applies.

Equation (6.47) describes an ideal MZ modulator. In real modulators there are many defects such as the insertion losses of the Y-branch couplers, propagation losses, unbalanced initial phase and amplitude as well as imbalances in the n_{eff} and the length L_p of the two arms. Betts [15] showed that all those defects will lead to a modification of Eq. (6.47) as

$$\frac{I}{\eta_{ins}I_o} = T(V) = \frac{1}{2}\left[(1 + E) + (1 - E)\cos\left(\phi_o + \frac{\pi}{V_\pi}V\right)\right], \tag{6.49}$$

where E is the ratio of minimum $I/\eta_{ins}I_o$ to maximum $I/\eta_{ins}I_o$ (also known as the contrast ratio), and ϕ_o is any residual phase imbalance in the two arms. Note that η_{ins} is the insertion efficiency ($10\log_{10}(\eta_{ins})$ is the insertion loss in dB), $T(V)$ is the intensity transmission normalized to the maximum output intensity at $\phi_o + \frac{\pi}{V_\pi}V = 0$ or $2n\pi$, and V is the instantaneous voltage applied to the electrode. For a well made LiNbO$_3$ MZ modulator, the insertion loss, including coupling losses to single mode optical fibers at the input and the output, is less than 3 dB. The value of E is 0.01 or less and $V_\pi L_p$ is of the order of 100 V mm at $\lambda=1.55$ μm.

The MZ modulation can also be achieved with any ratio of unbalanced $\Delta\phi$ in the two arms. The power output is zero whenever the difference of the total $\Delta\phi$ of the two arms is $n\pi$. However the phase of the output guided wave as a function of the applied voltage depends on the $\Delta\phi$ values.

Mach–Zehnder modulators are simply two phase modulators connected to two Y-branch couplers. Therefore, Eq. (6.34) to (6.41) are equally applicable to MZ modulators made in polymer and III-V compound semiconductor materials. However, the relation between Δn and $\boldsymbol{F}$ and $f_F(x,y)$ will differ in each case, depending on the material, waveguide and electrode configurations. *Section 3.1.2 discussed some examples of the optical waveguide configurations and the Δn as a function of $\boldsymbol{F}$ in those configurations. The major advantage of polymer material is its large r_{33} coefficient. However, how to produce polymer materials that will have a low propagation loss, large r_{33} coefficient and high transition temperature is still being investigated. The major advantage of the MZ modulator in III-V compound semiconductors is the small d that can be obtained in a reverse biased p–i–n structure. The disadvantage is that Y-branch and waveguides in III-V compound semiconductor materials usually have large losses due to scattering, contributing to a high insertion loss, and also r_{41} is smaller than the r_{33} in $LiNbO_3$ and polymers. In short, the $\Delta\phi$ in a MZ modulator could be created by any Δn. For example, in Section 3.3, the Δn produced by the electro-refraction ER effect in semiconductor materials was discussed. Also Δn could be obtained in materials that have a Kerr effect.*

The performance of MZ modulators will be gauged by both their optical and electrical characteristics. In analog applications, V consists of a bias voltage V_b and a RF instantaneous signal voltage v_{RF} which is much smaller than V_b. Therefore the $T(V)$ in Eq. (6.49) can be expressed in a Taylor's series expansion about $V = V_b$ as follows:

$$T(V_b + v_{RF}) = c_0 + \sum_{n=1}^{n=\infty} c_n v_{RF}^n, \tag{6.50}$$

where

$$c_0 = T(V_b) \quad \text{and} \quad c_n = \frac{1}{n!}\frac{d^n T}{dV^n}\bigg|_{V=V_b}. \tag{6.51}$$

Since T is a cosine function of V, we obtain

$$c_0 = \frac{1}{2}\left[1 + \cos\left(\phi_0 + \frac{\pi}{V_\pi}V_b\right)\right], \tag{6.52a}$$

$$c_n = \frac{1}{2}\frac{1}{n!}\left[\frac{\pi}{V_\pi}\right]^n (-1)^{\frac{n+1}{2}}\sin\left(\phi_0 + \frac{\pi}{V_\pi}V_b\right), \quad \text{for} \quad \text{odd } n \tag{6.52b}$$

$$c_n = \frac{1}{2}\frac{1}{n!}\left[\frac{\pi}{V_\pi}\right]^n (-1)^{\frac{n}{2}}\cos\left(\phi_0 + \frac{\pi}{V_\pi}V_b\right). \quad \text{for} \quad \text{even } n \tag{6.52c}$$

In analog applications, the output from the modulator is transmitted via optical fiber and detected. A linear photo-detector converts all the transmitted fraction of the output intensity $\eta_{ins}I_oT(V_b+v_{RF})$ into electrical currents proportional to c_0 and $c_n v_{RF}^n$. For a single frequency CW RF signal, $v_{RF} = A_{RF}\cos\omega_m t$, the photocurrent component at frequency ω_m (which is proportional to A_{RF}) is the output RF signal current. For a multiple frequency signal, the signal is proportional to $v_{RF} = \sum_j A_{RF,j}\cos\omega_{m,j}t$. The

signal current is proportional to c_1 in Eq. (6.52). *The output RF signal power is generated* *in the load of the photo-detector circuit by the RF signal current. In order to maximize the RF signal, V_b is set at $V_\pi/2$. The smaller the V_π, the more efficient is transmission of the RF signal. In addition, the RF signal power is also proportional to $(\eta_{ins}I_o)^2$. Note that the larger the I_o and the larger the η_{ins}, the larger is the RF signal.* On the other hand there are also distortions produced from higher orders of $T(V)$ that vary like $c_n v_{RF}^n$. At the bias voltage setting where $V_b = V_\pi/2$, c_n(for even n) is zero. The non-linear distortions of the v_{RF} are contributed mainly from c_n(for odd n)v_{RF}^n. Since the c_n coefficients are smaller at larger n and the nth order distortion is proportional to the nth power of v_{RF}, the distortions are negligible at small v_{RF} and increase faster than the signal at large v_{RF}. At very large v_{RF}, the distortions become as large as the signal. *The Spurious Free Dynamic Range (SFDR) of an analog link is the range of the RF signal power that may vary from minimum when the detected RF signal power is just equal to noise power to maximum when the power in the non-linear distortion terms becomes equal to the signal power.* Although the relations between the c_n coefficients are fixed for a given V_π in a simple MZ modulator which was described in Eq. (6.52), non-linear distortions can be reduced by various cancellation techniques such as pre-distortion, feed-forward and balanced modulators. Much research has been conducted to linearize analog links using MZ modulators and to maximize the SFDR [see 15].

The performance of MZ modulators in analog applications is measured by: (1) their η_{ins}, V_π and the maximum I_o before any saturation of c_1 occurs, (2) the SFDR of the link, and (3) the bandwidth of the link which is determined primarily by the capacitance of the modulator C_m.[17] *The C_m is the total capacitance of the phase modulator in the two arms of the MZ.*

In digital applications, MZ modulators are used to provide on and off switching of the optical signal. In that case the contrast ratio E is an important performance criterion in addition to V_π. However, phase variation associated with the modulation of the transmitting optical wave will cause pulse broadening in long distance transmission in optical fibers because of fiber dispersion. Therefore, performance of modulators is also measured by their chirp parameter, $\alpha = 2I\delta\phi/\delta I$ [16].[18] Gnauck et al. have shown that the chirp in a MZ modulator can be adjusted from -2 to $+2$ by varying the $\Delta\phi$ in the two arms of the MZ modulator to reduce the dispersion penalty [17]. The chirp is zero for a balanced MZ modulator in push-pull operation. Similarly to analog applications, the bandwidth of the MZ modulator in digital applications is also limited by its capacitance C_m.

No matter whether the MZ modulator is used in analog or digital links, the lower the V_π and the higher the η_{ins}, the more efficient is the modulation. According to Eq. (6.48), only $V_\pi L_p$ is determined by the material, waveguide and electrode configuration design of a specific modulator.[19] *For a given design, the longer the L_p, the lower is the V_π. However, like the phase modulators, there is also a trade-off between L_p and the electrical bandwidth. Since the MZ modulator is basically made from two phase modulators, the electrical frequency response of the link will be limited by the RC_m.*[20] *It means that the bandwidth is proportional to d and $1/L_p$. The longer the L_p, the smaller the electrical bandwidth. The smaller the d, the lower is the $V_\pi L_p$, and the smaller is the electrical bandwidth.*

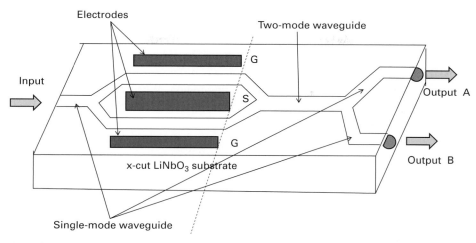

Fig. 6.5. A Mach–Zehnder switch on x-cut LiNbO$_3$ substrate. In this example, push-pull modulation is used, creating $+\Delta\phi$ and $-\Delta\phi$ in the two arms. G is the ground electrode and S is the signal electrode.

When the output Y-branch of the MZ modulator is replaced by a two-mode waveguide power divider as discussed in Section 6.1.1.4, the device becomes a switch. It is illustrated in Fig. 6.5 in a push-pull electrode configuration on an x-cut LiNbO$_3$ substrate. Its operation can be analyzed as follows. Following the electro-optic phase shift $\Delta\phi$ in the two arms of the balanced MZ in the push-pull mode, the amplitudes and phases of the guided waves incident on the output Y-branch power divider are

$A_{\text{in}} = \dfrac{A_{\text{o}}}{\sqrt{2}} e^{j\Delta\phi}$ and $B_{\text{in}} = \dfrac{A_{\text{o}}}{\sqrt{2}} e^{-j\Delta\phi}$. According to Eq. (6.8) and (6.9), the amplitudes

of the guided-wave in output waveguides A and B of the Y-branch power divider for such an input are

$$A_{\text{out}}(z = +L_{\text{o}}) = \frac{A_{\text{o}}}{\sqrt{2}} \cos\Delta\phi\, e^{-j\phi_{\text{s,in}}} e^{-j\phi_{\text{s,out}}} e^{-j2\beta_{\text{s}}L_{\text{c}}}$$
$$+ j\frac{A_{\text{o}}}{\sqrt{2}} \sin\Delta\phi\, e^{-j\phi_{\text{a,in}}} e^{-j\phi_{\text{a,out}}} e^{-j2\beta_{\text{a}}L_{\text{c}}}, \qquad (6.53)$$

$$B_{\text{out}}(z = +L_{\text{o}}) = \frac{A_{\text{o}}}{\sqrt{2}} \cos\Delta\phi\, e^{-j\phi_{\text{s,in}}} e^{-j\phi_{\text{s,out}}} e^{-j2\beta_{\text{s}}L_{\text{c}}}$$
$$- j\frac{A_{\text{o}}}{\sqrt{2}} \sin\Delta\phi\, e^{-j\phi_{\text{a,in}}} e^{-j\phi_{\text{a,out}}} e^{-j2\beta_{\text{a}}L_{\text{c}}}. \qquad (6.54)$$

Let the Y-branch power divider be designed so that $\Delta\phi_{\text{Y}} = (\phi_{\text{s,in}} + \phi_{\text{s,out}} + 2\beta_{\text{s}}L_{\text{c}}) - (\phi_{\text{a,in}} + \phi_{\text{a,out}} + 2\beta_{\text{a}}L_{\text{c}}) = \pi/2$. When $\cos\Delta\phi = \sin\Delta\phi$ (i.e. $\Delta\phi = \pi/4$), $A_{\text{out}} = 0$ and $|B_{\text{out}}| = |A_{\text{o}}|$. When $\cos\Delta\phi = -\sin\Delta\phi$ (i.e. $\Delta\phi = -\pi/4$), $|A_{\text{out}}| = |A_{\text{o}}|$ and $B_{\text{out}} = 0$. When $\Delta\phi = 0$, $\pi/2$, π, $3\pi/2$, etc., $|A_{\text{out}}| = |B_{\text{out}}| = \left|\dfrac{A_{\text{o}}}{\sqrt{2}}\right|$. Other ratios of $|A_{\text{out}}/B_{\text{out}}|$ are obtained when another $\Delta\phi$ is obtained. Clearly the trade-off between V_{π}, d, L_{p} and bandwidth for the MZ modulators applies also to the MZ switch.

6.2.1.3 The mode extinction modulator

The mode extinction modulator is one of the simplest modulators. Its principle of operation can be explained as follows. In order for a guided-wave mode to propagate effectively there must be a sufficient difference between its effective index and the indices of the substrate and cladding region. When the n_{eff} is reduced by the electro-optical effect so that it becomes close to, or lower than, the equivalent index of the cladding or substrate radiation modes, the guided-wave mode is close to or beyond cut-off. The power in the guided mode will then be coupled strongly to the radiation modes, and the guided-wave mode will have large propagation loss. The modulation of the propagation loss by the applied voltage produces a modulation of the optical power transmitted to the output of the waveguides. Since it is difficult to channel the radiated power effectively to another waveguide or location, it is just a modulator and not a switch. *In short, a mode extinction modulator is similar to a phase modulator as illustrated in Fig. 6.3, which has an n_{eff} of the waveguide mode close to cut-off. The effective index of the guided-wave mode is reduced by applying a voltage to the two electrodes parallel to the waveguide to create a $-\Delta n$. The intended effect of the $-\Delta n$ is to create propagation loss by the cut-off condition of the mode.*

Despite its simplicity, the mode extinction modulator has not been used much in applications for three reasons.

(1) The $-\Delta n$ that can be obtained by electro-optical effect is limited. For example, in LiNbO$_3$ (using the r_{33} as the electro-optic coefficient) and according to Eq. (3.19), when 100 volts is applied to a pair of electrodes separated by a 10 μm gap, the Δn_e is only approximately 1.6×10^{-4}.

(2) In order for the mode extinction mechanism to be effective for such a small Δn, the waveguide needs to have a small difference between the n_{eff} and the index of the cladding region in the absence of the applied electric field, much smaller than 0.001. In such a weak waveguide, the guided-wave energy is easily scattered into radiation modes by defects in the material and at the interfaces, resulting in a high propagation loss in the absence of the applied voltage.

(3) When a weak waveguide is excited experimentally from an external source such as a fiber, radiation modes are easily excited, resulting in poor coupling efficiency and large insertion loss.

Nevertheless, the concept of electro-optical loss modulation is very useful. In the section below, we discuss a similar approach, the electro-absorption (EA) modulator which is much more effective: EA modulators have already been developed into commercial products.

6.2.1.4 The electro-absorption modulator

The electro-absorption (EA) effect was discussed in Section 3.2. It is a modulation of the absorption of a well guided mode in III-V compound semiconductor waveguides. The increase in absorption is produced by an applied electric field for radiation at wavelength just below the bandgap of the material. The change in absorption coefficient may be

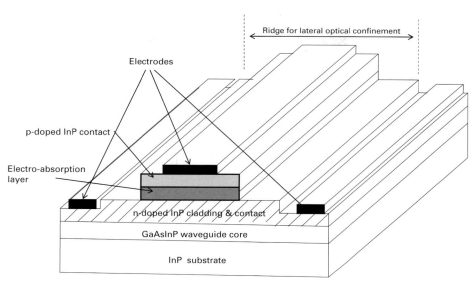

Ridge for lateral optical confinement

Electrodes

p-doped InP contact

Electro-absorption layer

n-doped InP cladding & contact

GaAsInP waveguide core

InP substrate

Fig. 6.6. An electro-absorption modulator.

caused by the Frantz–Keldysh effect in bulk semiconductors or by the Quantum Confined Stark Effect (QCSE) in quantum well materials. As we have shown in Eq. (6.40), the propagation of the mth guided-wave mode in a channel waveguide can be described by

$$E(x, y, z) = A\boldsymbol{e_m}(\boldsymbol{x}, \boldsymbol{y})e^{-\frac{\alpha_m}{2}z}e^{-jn_{\text{eff},m}kz}, \qquad (6.55)$$

where $\alpha_m = \alpha_{m,o} + \Delta\alpha_m$. $\alpha_{m,o}$ is the residual absorption coefficient of the intensity of the mth guided-wave mode in the absence of applied electric field, and $\Delta\alpha_m$ is the change of the absorption coefficient created by the $\Delta\alpha$ in the electro-absorption medium by the instantaneous electric field F.[21] No matter what is the mechanism creating it, the $\Delta\alpha_m$ creates a loss modulation of the guided wave propagating in the waveguide. In contrast to extinction modulators, electro-absorption modulation is applied to a waveguide that has a well guided mode which has low residual propagation loss and high coupling efficiency to fibers (or other waveguides). Figure 6.6 illustrates an EA waveguide modulator. It shows a ridged waveguide on InP substrate, where the waveguide core consists of a quaternary InGaAsP layer sandwiched between lower index InP layers in the vertical direction. A ridge is etched on the top cladding layer to provide the mode confinement in the lateral direction. Within the cladding layers above the core, there is an EA layer. In order to provide a large electric field from a given applied voltage, the EA layer is the intrinsic layer of a reverse biased p–i–n diode. The doped p and n layers are in contact with the metallic electrodes.

When there is an applied electric field F, the absorption coefficient of the EA material can be described as $\alpha = \alpha(F)$.[22] Figure 6.7 shows typical measured absorption spectra α in cm^{-1} as a function of wavelength at various applied voltages.[23] This sample was designed for EA operation at the 1.3 μm radiation wavelength. The voltage is applied across a 0.14 μm thickness intrinsic absorption layer which consists of eight periods of

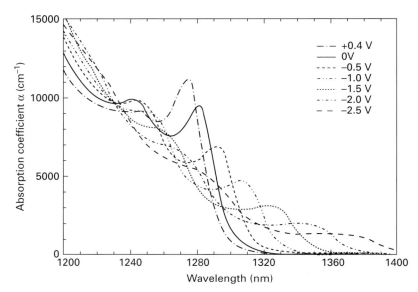

Fig. 6.7. Measured absorption coefficient of an InAsP/GaInP multiple quantum well material as a function of wavelength at various reverse biased voltages. (This figure is taken from ref. 18 with permission from K. K. Loi.)

8.9 nm compressively strained InAsP quantum wells and 7.4 nm tensile strained GaInP barriers. Since the voltage drops across the p and n layers are negligible, the electric field F in the EA material is approximately $V(1.4 \times 10^{-6})$ volts/meter. Note that α_{o} and $\Delta\alpha$ at the 1.3 μm wavelength as a function of F can be obtained directly from the measured absorption spectra.

From the perturbation analysis, the α_m of the mth guided-wave mode is related to α of the absorption material by

$$\Delta\alpha_m = \Gamma\Delta\alpha_{\mathrm{av}} = \frac{\displaystyle\int_{\mathrm{EA\ material\ region}} \Delta\alpha\underline{e_m} \bullet \underline{e_m^*}\mathrm{d}x\,\mathrm{d}y}{\displaystyle\int \underline{e_m} \bullet \underline{e_m^*}\mathrm{d}x\,\mathrm{d}y}. \tag{6.56}$$

For uniform α in the EA material, $\Delta\alpha_{\mathrm{av}} = \Delta\alpha(F)$ and

$$\Gamma = \frac{\displaystyle\int_{\mathrm{EA\ material\ region}} \underline{e_m} \bullet \underline{e_m^*}\mathrm{d}x\,\mathrm{d}y}{\displaystyle\int \underline{e_m} \bullet \underline{e_m^*}\mathrm{d}x\,\mathrm{d}y} \tag{6.57}$$

where Γ is known as the overlap integral of the EA material. It is the fraction of the optical energy of the guided-wave mode in the absorption layer. The power I carried by this mode at the output of the EA modulator L long is

$$\frac{I}{I_{\mathrm{o}}\mathrm{e}^{-\alpha_{m,\mathrm{o}}L}} = T(F) = \mathrm{e}^{-\Gamma\Delta\alpha_{\mathrm{av}}(F)L}, \tag{6.58}$$

where I_{o} is the incident optical power of the mth mode, and $T(F)$ is the power transmission of the modulator, normalized with respect to the transmission when $F = 0$. Note that

$T(F) = 1$ at $F = 0$. Like the electrodes used in phase and MZ modulators, the electric field F created from an applied electric voltage V can again be expressed as

$$\underline{F} = \frac{V}{d} \underline{f}_F(x, y).$$

For multiple quantum well materials, F is in the direction perpendicular to the EA layer. Typically V consists of a bias voltage V_b plus a RF signal voltage V_{RF}. Similar to Eq. (6.50) and (6.51), T can also be expressed by a Taylor's series expansion about $T(V_b)$,

$$T(V_b + v_{RF}) = c_0 + \sum_{n=1}^{\infty} c_n v_{RF}^n,$$

$$c_0 = T(V_b), \qquad c_n = \frac{1}{n!} \frac{\partial^n T}{\partial V^n}\bigg|_{V=V_b}.$$

In analog applications, the output from the modulator is transmitted via an optical fiber and detected. A linear photo-detector converts all the transmitted output intensity $e^{-\alpha_{m,o}L} I_o T(V_b + v_{RF})$ into electrical currents. The photo-current component proportional to $c_1 v_{RF}$ produces the output RF signal power in the detector circuit. The RF signal power transmitted to the load is proportional to

$$[I_o e^{-\alpha_{m,o}} c_1 v_{RF}]^2 = \left[I_o e^{-\alpha_{m,o}L} \frac{\partial T}{\partial V}\bigg|_{V_b} v_{RF} \right]^2$$

$$= \left[-I_o \Gamma e^{-\alpha_{m,o}L} e^{-\Gamma \Delta \alpha_{av}|_{V_b} L} L \frac{\partial \Delta \alpha_{av}}{\partial V}\bigg|_{v_b} \right]^2. \tag{6.59}$$

The detected RF signal power can be increased in several ways.

(1) *It can be increased by designing and fabricating an EA material, waveguide and electrode structure such that* $\dfrac{\partial \Delta \alpha_{av}}{\partial V}\bigg|_{V_b} = \dfrac{\partial \Delta \alpha_{av}}{\partial F}\bigg|_{V_b} \bullet \dfrac{\partial F}{\partial V}$ *is large. From Fig. 6.7, it is clear that multiple quantum well materials that have the sharpest exciton absorption peak, largest quantum confined Stark shift and appropriate separation of the exciton absorption peak from the radiation wavelength will yield the largest* $\dfrac{\partial \alpha_{av}}{\partial F}\bigg|_{V_b}$.

(2) *From Eq. (6.44), it is clear that the p–i–n structure and the electrode design determine the* $F(x, y)/V = f_F(x, y)/d$ *in the EA medium.*

(3) *The detected RF signal can be increased by increasing* η_{ins} *and the optical power* I_o.

(4) Γ *depends on the fraction of the guided-wave energy in the EA medium. For a given electrode, waveguide and p–i–n structure configurations, the RF signal is maximized by choosing an L that maximizes* $\Gamma L \exp\left[-\left(\alpha_{m,o} + \Gamma \Delta \alpha_{av}|_{V_b} \right) L \right]$.

(5) *For a given modulator, the RF signal is maximized by using a* V_b *which maximizes* $\dfrac{\partial T}{\partial V}\bigg|_{V_b}$.

Note that the RF signal power is proportional to $(c_1 v_{RF})^2$ in both MZ and EA modulators. It is helpful to compare the effectiveness of two different types of modulator in generating the RF signal (i.e. $c_1 v_{RF}$) as follows. Let there be a virtual equivalent MZ modulator biased at the V_b so that $\phi_o + \dfrac{\pi}{V_\pi} V_b$ is $\pi/2$. The T_{MZ} of the equivalent MZ modulator provides the same $c_1 v_{RF}$ as the T_{EA} of a given EA modulator. Then, according to Eq. (6.51), this virtual MZ modulator should have a $V_{\pi,eq}$ such that,

$$V_{\pi,eq} = \frac{\pi}{2} \left[\left. \frac{\partial T_{EA}}{\partial V} \right|_{V_b} \right]^{-1}. \tag{6.60}$$

In other words, for each EA modulator there is a virtual equivalent MZ modulator with a $V_{\pi,eq}$ that will yield the same $c_1 v_{RF}$. In the literature, $V_{\pi,eq}$ is often used to designate the modulation efficiency of EA modulators. Loi *et al.* have demonstrated a $V_{\pi,e} \approx 2$ V in an InAsP/GaInP multiple quantum well modulator 185 µm long at the 1.3 µm wavelength [19]. This device had an 11 GHz bandwidth. Another similar modulator 90 µm long had an electrical capacitance of 0.22 pF and a measured bandwidth over 20 GHz.

Similar to the discussion in Section 6.2.1.2 for MZ modulators, the SFDR of a photonic link will be limited by the noise, the maximum I_o allowed without saturation of $d\alpha/dV$ and the distortions generated from the $c_n v_{RF}^n$ terms of the $T(V_b + v_{RF})$. However, different than the MZ modulators, there is no V_b setting that will make $c_2 = 0$. There may be other methods that may be used to reduce or to eliminate the effect of specific order of distortion. For example, for analog applications requiring only sub-octave bandwidth, electrical filters may be used to remove the frequency components produced by different orders of distortions outside the pass band of the filter. However, whenever v_{RF} consists of more than one RF signal frequency, there are still distortions produced by mixing of these RF signals in the higher order $c_n v_{RF}^n$ terms. Some of these distortions, called inter-modulations (IM), have frequencies that fall within the pass band of the electrical filter. Betts gives a detailed listing of the magnitude of the distortion signals at various frequencies when there are two inputs, $v_{RF} = v_a \sin \omega_a t + v_b \sin \omega_b t$, including all $c_n v_{RF}^n$ up to $c_5 v_{RF}^5$ [15]. For sub-octave applications with an electrical filter, the non-linear distortions due to c_2 and c_4 are filtered out. Note that V_b is set so that the third order distortion, is minimized. For multi-octave applications, V_b may be set to minimize c_2. In an InGaAsP multiple quantum well EA modulator for the 1.55 µm radiation wavelength, Zhuang has demonstrated a SFDR of 132 dB-Hz$^{4/5}$ for sub-octave applications with the spurious signal dominated by the fifth order distortion, and a SFDR of 118 dB-Hz$^{2/3}$ for multi-octave applications with the spurious signal dominated by the third order distortion [20].[24]

A particular characteristic of the EA modulator that does not apply to MZ electro-optic modulators is the saturation of $c_1 v_{RF}$ of the EA modulation at very large I_o. Saturation in an EA modulator is caused primarily by the photo-current generated from the absorbed radiation. The photo-current creates three effects.

(1) *High concentration of photo-generated carriers in the i layer shields the applied electric field, thereby reducing the absorption. Methods such as peripheral coupling may be used to reduce the saturation effect for a given I_o [21].*

(2) *The RF photo-current passes through the same electrical circuit as the RF signal source driving the modulator. It creates a negative feedback of the voltage applied to the modulator from the RF source.*

(3) *The photo-current reduces the junction resistance of the p–i–n junction significantly so that the modulator impedance is no longer matched to the RF source. Although the effect of negative feedback and impedance change is minor at low and moderate I_o, the output RF signal is no longer proportional to I_o^2 at very large I_o.*

The ratio of the output RF power to the RF signal power in the photonic link is completely saturated at high optical power [22].

For digital applications, the important optical modulator characteristics are the contrast ratio that can be achieved for a given applied voltage and the frequency chirp parameter α. The optical power handled by the modulator is typically a few dBm. Thus the saturation property of the modulator is usually not important. (1) The EA modulator performance is rated by the V required to achieve a specific contrast ratio. From Eq. (6.58), it is clear that the larger the applied V, the larger is the contrast ratio. When an EA modulator is coupled to other components such as fibers at the input and output ends, there may be radiation coupled to the output through optical leakage paths that limits the maximum obtainable contrast ratio. (2) Comparing with the electro-optic MZ modulators, the optical chirping characteristics of the EA modulators are not as desirable for long distance fiber transmission. In Section 3.3, the electro-refraction effect was discussed. In any material, the refractive index spectrum is always related to the absorption spectrum by the Kramers–Kronig relationship shown in Eq. (3.27). Therefore, there is always a change of the phase of the output guided wave as a function of change of absorption in EA modulators. Fells et al. measured the change of Δn in EA modulators as a function of Δk which is the imaginary part of the complex refractive index representing absorption [23]. Devaux et al. showed that the α of an InGaAsP/InGaAsP multiple quantum well EA modulator varied from 0.5 to 2.5, depending on the bias voltage [24].

The parallel electrodes and the p–i–n structure may be represented by an electrical impedance which consists of a capacitance C_m in parallel with a junction resistance R_j. Therefore, the electrical circuit bandwidth of the EA modulator will be limited again by the $1/R_sC_m$. There is a trade-off between optical and electrical performance. The longer the device length L, the more efficient is the modulator, and the lower is the bandwidth. In comparison, the EA modulator is much shorter than the MZ electro-optic modulator. Therefore, EA modulators could have a large bandwidth by using a small C_m. Ido et al. have demonstrated a high-speed InGaAs/InAlAs modulator integrated with InGaAsP/InP passive waveguides by minimizing the C_m at the 1.55 μm optical wavelength [25]. Their 150 μm long device has 21 GHz bandwidth, while their 50 μm long device has 33 GHz bandwidth. In the digital mode the driving voltage is 3 V and the insertion loss is 8 dB. The optical input power is +5 dBm. *When EA modulators are compared with electro-optic MZ modulators, their biggest advantages are the low $V_{\pi,eq}$ and the length of the device. For the same $V_{\pi,eq}$ their length is typically a fraction of a millimeter or less, while the MZ modulators are centimeters long. The shortness of the EA device is also attractive physically for many applications. Since the EA effect is used for loss modulation, it has*

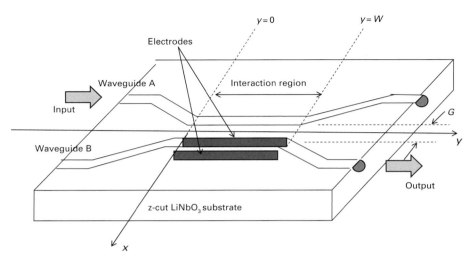

Fig. 6.8.   A directional coupler modulator on z-cut LiNbO₃ substrate.

not been utilized for switching. The semiconductor EA modulators can be monolithically integrated with lasers or other electronic components on the same chip. Kawanishi et al. have demonstrated an EAM-DFB integrated laser for digital applications with 40 GHz bandwidth and 12 dB contrast ratio [26].

6.2.1.5 Directional coupler modulators and switches

When a directional coupler power divider as discussed in Section 6.1.1.2 is made from electro-optic material an electrode could be used to apply an electric field on one of the waveguides as shown in Fig. 6.8 for a directional coupler modulator on z-cut LiNbO₃. In most directional couplers, waveguide A and waveguide B are identical single mode waveguides. They are well isolated from each other outside the interaction region. Within the interaction region they are separated by a gap G. The input power is incident on waveguide A. Typically, a directional coupler modulator will transfer 100% of the optical power from the guided-wave in the input waveguide A to the output waveguide B in the absence of the applied voltage. When a voltage is applied to waveguide B, the $\Delta\beta$ reduces the power transferred from A to B, thereby producing a modulation of the output power of waveguide B. An analysis of the directional coupler modulator is presented below.

Similarly to the phase modulator discussed in Section 6.2.1.1, the instantaneous applied voltage creates a change of the refractive index Δn of the electro-optic material which produces a change of the effective index of the guided-wave mode $\Delta n_{\text{eff},m}$ in waveguide B in the interaction region. From the perturbation analysis, Eq. (3.38) and Eq. (3.39), we obtain

$$\Delta\beta = \Delta n_{\text{eff},m}k = k\frac{n}{n_{\text{eff},m}}\Gamma\Delta n_{\text{av}} = k\frac{n}{n_{\text{eff},m}}\frac{\displaystyle\iint_{\text{electro-optic region}}\Delta n\underline{e_m}\bullet\underline{e_m^*}\,dx\,dy}{\displaystyle\int_{-\infty}^{\infty}\int_{-\infty}^{\infty}\underline{e_m}\bullet\underline{e_m^*}\,dx\,dy}. \tag{6.61}$$

When $\Delta\beta$ is produced by the linear electro-optic effect, $\Delta\beta$ is proportional to F. Since $F = Vf_F/d$, $\Delta\beta$ is also proportional to the instantaneous applied voltage V. In the lumped element approximation, the time dependence of $\Delta\beta$ on the time variation of F is independent of z.

Let the interaction region length of the symmetric directional coupler be W, from $y = 0$ to $y = W$. When waveguides A and B are placed close to each other with a sufficiently small gap G, the propagation of the guided-wave mode in waveguides A and B is coupled, with the coupling coefficient C given in Eq. (2.21). The guided-wave amplitudes a_A and a_B for an incident wave with amplitude A in waveguide A are given in Eq. (2.22) as follows

$$a_A = Ae^{j\frac{\Delta\beta}{2}z}\left[\cos\left(\sqrt{C^2 + (\Delta\beta/2)^2}\,y\right)\right.$$

$$\left. -j\frac{(\Delta\beta/2)}{\sqrt{C^2 + (\Delta\beta/2)^2}}\sin\left(\sqrt{C^2 + (\Delta\beta/2)^2}\,y\right)\right],$$

$$a_B = \frac{-jCA}{\sqrt{C^2 + (\Delta\beta/2)^2}}e^{-j\frac{\Delta\beta}{2}z}\sin\left(\sqrt{C^2 + (\Delta\beta/2)^2}\,y\right). \tag{6.62}$$

For symmetrical waveguides and in the absence of applied electric field, $C_{AB} = C_{BA} = C$ and $\beta_a = \beta_b$ (i.e. $\Delta\beta = 0$). When W is chosen so that $CW = \pi/2$, then, $\left|\frac{a_B}{A}\right| = 1$ and $I_{out}/I_{in} = 1$. As a voltage V is applied to the electrode, a_B decreases. At $V = V_\pi$ the $\Delta\beta$ is so large that $C^2 + \left(\frac{\Delta\beta}{2}\right)^2 = \left(\frac{\pi}{W}\right)^2$, then $a_B = 0$. Note that V_π is defined here similarly to the V_π of the MZ modulator, it is the voltage at which $I_{out} = 0$. From Eq. (6.62)

$$\Delta\beta|_{V=V_\pi} \bullet W = \sqrt{3}\pi. \tag{6.63}$$

Since $\Delta\beta = \frac{V}{V_\pi}\Delta\beta|_{V=V_\pi}$, we obtain

$$\frac{I_{out}}{\eta_{ins}I_{in}} = T(V) = \frac{1}{1 + 3[V/V_\pi]^2}\sin^2\left[\frac{\pi}{2}\sqrt{1 + 3(V/V_\pi)^2}\right], \tag{6.64}$$

where the insertion efficiency η_{ins} is added to take care of any residual propagation loss, and V is the instantaneous voltage applied to the electrode. If a second electrode is also applied to waveguide A to create a $-\Delta\beta$ in the push-pull operational mode, then $\Delta\beta = 2\frac{V}{V_\pi}\Delta\beta|_{V=V_\pi}$ and (V/V_π) is replaced by $(2V/V_\pi)$ in Eq. (6.64). Just like the MZ electro-optic modulators, V_π will depend on the material, the waveguide configuration and the electrode design. For LiNbO$_3$ waveguides, we obtain

$$V_\pi W = \frac{\sqrt{3}\lambda dn_{eff,m}}{n_e^4 r_{33}}\left[\frac{\iint f_F(x,y)\underline{e}\bullet\underline{e}^*dx\,dy}{\iint \underline{e}\bullet\underline{e}^*dx\,dy}\right]^{-1}. \tag{6.65}$$

Comparing Eq. (6.64) with Eq. (6.49), we see that the directional coupler (DC) modulator is very similar to the MZ modulator/switch. The transfer function T(V) of both types of modulator has a sinusoidal dependence on V/V_π. The electrode and waveguide *configurations are similar, yielding the same dependence of V_π and capacitance C_m on f_F, d, $\underline{e_m}$, r_{ij}, and n_{eff}. There is a trade-off between V_π and W (or L_p) as well as between the RC bandwidth and V_π. The electrical bandwidth of both types of modulator is limited by C_m. The more subtle differences are: (1) the T(V) for a directional coupler has a damped sinusoidal dependence. (2) For analog applications there is no bias voltage at which the second order term d^2T/dV^2 of the directional coupler modulator is zero. (3) For digital applications there is no setting of the directional coupler such that the chirping parameter α can be adjusted like a MZ modulator. (4) For directional couplers, $CW=\pi/2$. In order to decrease V_π by increasing W, C needs to be reduced by increasing the gap of separation between the optical waveguides in the coupling region. At large separation, the coupling coefficient C becomes more sensitive to wavelength variation and fabrication error. Therefore it is difficult to make a practical long directional coupler modulator with low V_π in materials such as LiNbO$_3$. (5) Directional coupler modulators are automatically switches.*

6.2.2   Traveling wave modulators and switches

When the wavelength of the electrical signal is comparable to, or shorter than, the length of the electrode structure, the voltage and current on the electrodes that provide the electric field for electro-optical effects are both a function of position and time. The device can no longer be represented in the electrical circuit driving it by lumped circuit elements, R, L and C. The optical wave no longer experiences a time invariant electro-optical effect as it propagates through the device. In Section 4.2.1, we have shown that the electrical characteristics of the electrode structure can now be represented as an electrical transmission line. In addition, the ωL of the inductance of the electrode structure is now significant at high RF frequencies. The conductance of the metal in the electrode can no longer be regarded as ∞. The electrode is now represented by a transmission line from $z = 0$ to $z = L_p$ or W. Furthermore, the impedance representing the electrode to the RF driving circuit will depend on the location of the connection on the transmission line and the terminal impedances. There may be reflections of RF signals at the ends or at any discontinuity. The time varying electrical field on the electrode produced by the RF signal source will depend on how the electrodes are connected to the external circuits and how the transmission line is terminated.

In order to provide efficient modulation of the optical guided wave proportional to the time variation of the RF drive signal without the interference of the reflected RF signals, the electrical signal from the RF source is usually transmitted to the electrode through the beginning of the transitional transmission line at $z = 0$, and the electrode transmission line is terminated at the end by a matched load at $z = L$, L_p or W. Ideally, it is desirable to have the internal impedance of the RF source, the characteristic impedance of the transitional transmission line and the impedance of the load matched to each other so that there will not be reflections of electrical signals from any discontinuity. Since the source impedance is typically 50 Ω, it is desirable to have 50 Ω for the characteristic impedance for the transition and electrode transmission lines and for the impedance of the load.

For most of the electrodes used in modulators and switches, the fundamental electric mode of the transmission line is a TEM mode. For a CW microwave signal at frequency ω_m, the microwave electric field propagating in the $+z$ direction without any reflections can be expressed as

$$F_{RF}(x,y,z,t) = \frac{1}{2} F_{RF}^o(x,y) e^{-\frac{\alpha_{RF}}{2} z} \left[e^{j(\omega_m t - \beta_{RF} z)} + \text{complex conjugate} \right], \tag{6.66}$$

where $F_{RF}^o(x,y)$ is the transverse electric field, α_{RF} is the attenuation of the microwave intensity in the z direction, and $\beta_{RF} = \sqrt{\varepsilon_m/\varepsilon_o}\, \frac{\omega_m}{c} = n_m \omega_m/c$ is the microwave propagation wave number of the mode. Note that ε_m is the effective permittivity of the medium between electrodes at microwave frequency, and n_m is the equivalent index of the microwave mode. The microwave phase velocity is c/n_m. When a transmission line with matched load is connected to the RF source, it is represented by an impedance Z_o to the source. When Z_o is matched to the source impedance, the voltage applied to the input of the modulator at all frequencies is a constant $V_s/2$ where V_s is the RF source voltage. There is no longer an RC_m bandwidth limitation within which the modulator voltage is reduced to less than $1/\sqrt{2}$ of its peak value as in low frequency modulation.

6.2.2.1 Traveling wave electro-optic modulators

In electro-optic modulators or switches, the $F_{RF}(x,y,z,t)$ of the microwave signal at frequency ω_m produces an instantaneous Δn_{eff} of the optical guided-wave mode at z

$$\Delta n_{\text{eff,RF}}(z,t) = \frac{1}{2} \Delta n_{\text{eff}}^o e^{-\frac{\alpha_{RF}}{2} z} \left[e^{j(\omega_m t - \beta_{RF} z)} + \text{complex conjugate} \right]. \tag{6.67}$$

The relation between Δn_{eff}^o and F_{RF}^o is the same as those derived for low frequency devices in Eq. (6.42) and (6.61) for phase, MZ and directional coupler modulators.

(A) Traveling wave phase modulators

In phase modulators, the optical guided wave propagates in the z direction with propagation wave number $n_{\text{eff}}k$. The $\Delta n_{\text{eff}}(z,t)$ seen by the photons entering the waveguide in the mth mode at $z = 0$ is

$$\begin{aligned}
\Delta n_{\text{eff}} &= \Delta n_{\text{eff}}^o e^{-\frac{\alpha_{RF}}{2} z} \cos\left[\omega_m t - \left(n_m - n_{\text{eff},m} \right) kz \right] \\
&= \Delta n_{\text{eff}}^o e^{-\frac{\alpha_{RF}}{2} z} \cos(\omega_m t - \delta n k z),
\end{aligned} \tag{6.68}$$

where $\delta n = n_m - n_{\text{eff},m}$. The electric field of the mth mode of the guided-wave can be described as

$$\underline{E(x,y,z)} = A\mathbf{e}_m(x,y) e^{-\frac{\alpha_g}{2} z} e^{-jn_{\text{eff},m} z} e^{-j\left(\Delta n_{\text{eff},m}^o \int_0^z e^{-\frac{\alpha_{RF}}{2} z} \cos(\omega_m t - \delta n k z)\, dz \right)}. \tag{6.69}$$

Therefore, when α_{RF} is neglected, the total phase shift $\Delta\phi$ of the guided wave produced by the CW traveling wave RF signal for a distance of L_p is

$$\Delta\phi = \Delta n^{\circ}_{\text{eff},m} \int_{0}^{L_{\text{p}}} \cos(\omega_m t - \delta n k) \mathrm{d}z$$

$$= \Delta n^{\circ}_{\text{eff},m} L_{\text{p}} \frac{\sin(\delta n k L_{\text{p}}/2)}{(\delta n k L_{\text{p}}/2)} \cos(\omega_m t - \delta n k L_{\text{p}}). \tag{6.70}$$

Note that $\Delta\phi$ is directly applicable to phase modulators, and that when the microwave equivalent index n_m matches the optical effective index $n_{\text{eff},m}$, $\delta n = 0$, and the $\Delta\phi$ modulation is a constant at all RF frequencies for a constant voltage. Note that $\Delta\phi$ is sensitive to δn at large kL_{p}. When $\delta n k L_{\text{p}}/2 = 1.4$, $\Delta\phi$ is reduced to $1\sqrt{2}$ of its maximum value at $\omega = 0$.[25] Therefore, the bandwidth of traveling wave modulation for a given δn and L_{p} is given in the literature to be

$$L_{\text{p}}\Delta\omega_m = \frac{2.8c}{\delta n}. \tag{6.71}$$

As we have discussed earlier for low frequency operations, the longer the L_{p}, the smaller is the $\Delta n^{\circ}_{\text{eff},m}$ required to yield a given $\Delta\phi$, and the smaller is the required RF modulation voltage. There is no RC limitation of electrical bandwidth. Therefore traveling wave phase modulators can operate with low drive voltage and large bandwidth. However, α_{RF} is also very important in determining $\Delta\phi$. It increases as the RF frequency ω_m is increased.[26] Thus it limits both the effectiveness of using large L_{p} to reduce the drive voltage and the bandwidth of the modulator. In Section 4.4, the more general result including α_{RF} was derived.[27] For an electrode transmission line well matched to the RF source and the load, the bandwidth of the traveling wave phase modulator is limited by $\delta n L_{\text{p}}$ and α_{RF}. In practice, matching of the transmission line may not be good. There are reflections of microwaves. Then the interference effects produced by the reflections of microwaves created at various discontinuities further limit the bandwidth of the modulator.

(B) Traveling wave Mach–Zehnder modulators and switches

The traveling wave Mach–Zehnder modulator consists of essentially two Y-branch power dividers interconnected by two arms which are single mode waveguides. One or both of the arms are phase modulators. Since the $\Delta\phi$ of the phase modulator determines the intensity of the output, the characteristics of the MZ modulators at high frequency are determined by both the characteristics of phase modulation at high frequencies and the MZ modulator characteristics for a given $\Delta\phi$ at low frequencies. The V_{π} for modulators with long L_{p} and good $\Gamma\Delta n_{\text{av}}$ could be quite low. The bandwidth is determined by the $\delta n L_{\text{p}}$, α_{RF} and the matching of the electrode transmission line with the RF source and the load. The characteristics of the Mach–Zehnder switch are the same as the MZ modulator connected to a Y-branch power divider.

Although the principles of a traveling wave electro-optic MZ modulator are simple, the engineering design of the electrode and waveguide configurations that will yield low δn and α_{RF}, large $\Delta n_{\text{eff},m}$, long L_{p} and good impedance matching is very demanding. A fully packaged LiNbO$_3$ traveling wave modulator with 5 V and 40 GHz bandwidth, using

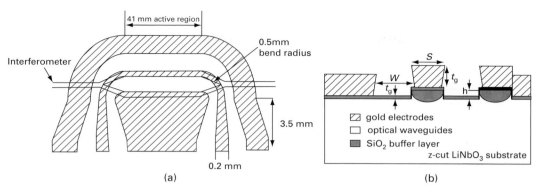

Fig. 6.9.   A fully packaged broadband Mach–Zehnder traveling wave LiNbO$_3$ modulator. (Taken from ref. 27 with permission of IEEE.) (a) Top view of the coplanar waveguide electrode on top of the optical MZ interferometer. (b) Cross-section of the electrodes and the ridged optical waveguides in the interferometer region. The ridge is used to improve the electro-optic overlap G. The thin SiO$_2$ layer is used to improve electrical conduction. The dimensions of the gold electrodes, $S = 8$ mm, $W = 25$ μm and $t_g = 21$ μm , are designed to minimize δn and to match the 50 Ω impedance of the RF source. The load and the RF source are not shown in the figure.

electrodes 41 mm long, was reported by Howerton *et al.* [27]. The V_π of this device at DC is 2.2 V. Figure 6.9 Illustrates their device. Broadband operation with 100 GHz bandwidth has been reported by Noguchi *et al.* [28].

(C) Traveling wave directional coupler modulators and switches

For directional couplers as illustrated in Fig. 6.8, the propagating RF voltage on the electrode transmission line created a propagating electro-optic $\Delta n_{\mathrm{eff}}(z,t)$ on waveguide B as shown in Eq. (6.67). For photons propagating from $z = 0$ to $z = W$ and for a CW RF signal at ω_m propagating in the $+z$ direction without reflection, it sees a $\Delta\beta$ in its propagation, i.e.

$$\beta_\mathrm{B} = n_{\mathrm{eff},m}k + \Delta\beta(z,t),$$

$$\Delta\beta = \Delta n_{\mathrm{eff},m}(z,t)k = k\frac{n}{n_{\mathrm{eff},m}}\Gamma\Delta n_{\mathrm{av}}$$

$$= k\frac{n}{n_{\mathrm{eff},m}}\frac{\displaystyle\iint_{\text{electro-optic region}}\Delta n e^{-\frac{\alpha_{\mathrm{RF}}}{2}z}\cos(\omega_m t - \delta nkz)\underline{e_m}\bullet\underline{e_m^*}\,\mathrm{d}x\,\mathrm{d}y}{\displaystyle\int_{-\infty}^{\infty}\int_{-\infty}^{\infty}\underline{e_m}\bullet\underline{e_m^*}\,\mathrm{d}x\,\mathrm{d}y}. \tag{6.72}$$

For directional couplers using identical waveguides A and B, the coupled mode equation shown in Eq. (2.21) will now be:

$$\frac{\mathrm{d}a_\mathrm{A}}{\mathrm{d}z} = -\mathrm{j}Ce^{\mathrm{j}\Delta\beta(z,t)z}a_\mathrm{B}(z),$$

$$\frac{\mathrm{d}a_\mathrm{B}}{\mathrm{d}z} = -\mathrm{j}Ce^{-\mathrm{j}\Delta\beta(z,t)z}a_\mathrm{A}(z). \tag{6.73}$$

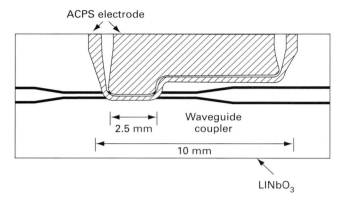

Fig. 6.10.   Top view of a traveling wave electro-optic directional coupler modulator. (Taken from ref. 29 with permission from American Institute of Physics.) The asymmetric coplanar strip transmission line for the electrodes has a 35 Ω characteristic impedance. It has a strip width of 15 mm, a gap of 5 μm and a thickness of 3–4 μm.

Note that $\Delta\beta$ is proportional to the applied voltage at $z = 0$, $V = V_{max} \cos \omega_m t$. The solution of Eq. (6.73) is complex for finite α_{RF} and δn. When $\alpha_{RF} = 0$ and $\delta n = 0$, $\Delta\beta$ is independent of z. The solution of Eq. (6.73) is identical to the result given in Eq. (6.62) for the low frequency (or very short) directional modulators. Results obtained in Eq. (6.63) to (6.65) are directly applicable. In principle, the V_π can be very low by using long W and small coupling coefficient C. There will be no limitation on the bandwidth caused by the C_m. A 1 cm long traveling wave z-cut LiNbO$_3$ directional coupler modulator with 2.5 mm long electrode, 22 GHz 3 dB electrical bandwidth and $V_\pi = 26$ V has been reported by Korotky *et al.* [29]. Figure 6.10 illustrates the top view of this modulator on z-cut LiNbO$_3$. Note that the ACPS electrode creates a push-pull operation in $\Delta\phi$. The gold electrodes have a characteristic impedance of 35 Ω. A tapered electrode is used to extend the electrodes to the edge of the LiNbO$_3$ to mate with external microwave circuits. The device bandwidth was limited by δn.

Because of the difficulties of fabricating devices with long W with precise control of small C, traveling wave directional modulators with low V_π have not been reported.

6.2.2.2   Traveling wave electro-absorption modulators

In electro-absorption EA modulators, the voltage applied to the electrode consists of a bias and a RF voltage, $V = V_b + V_{RF}(t)$. For a CW RF signal, $V_{RF}(t) = V_{RF}^o \cos \omega_m t$, applied to the transmission line of the electrodes of an EA modulator at $z = 0$ without reflection, the $F_{RF}(x,y,z,t)$ described in Eq. (6.66) creates a $\Delta\alpha_{RF}(x,y,z,t)$ within the EA medium given by

$$\Delta\alpha_{RF}(x, y, z, t) = \frac{1}{2}\Delta\alpha_{RF}^o(x, y)e^{-\frac{\alpha_{RF}}{2}z}\left[e^{j(\omega_m t - \beta_{RF}z)} + \text{complex conjugate}\right]. \quad (6.74)$$

The relation between F_{RF}^o and $\Delta\alpha_{RF}^o$ is the same as those discussed in Sections 3.2 and 6.2.1.4.[28] The $\Delta\alpha_{RF}$ within the EA medium creates a traveling wave $\Delta\alpha_{m,RF}$ propagating from $z = 0$ to $z = L$ in the optical waveguide, so that

$$\Delta\alpha_{m,\mathrm{RF}}(z, t) = \frac{1}{2}\Delta\alpha^{\mathrm{o}}_{m,\mathrm{RF}}\, e^{-\frac{\alpha_{\mathrm{RF}}}{2}z}\left[e^{\mathrm{j}(\omega_m t - \beta_{\mathrm{RF}}z)} + \text{complex conjugate}\right], \tag{6.75}$$

where the relation between $\Delta\alpha_{m,\mathrm{RF}}$ and $\Delta\alpha^{\mathrm{o}}_{\mathrm{RF}}$ is the same as that given in Eq. (6.56) for low frequency EA modulators. Photons propagate in the mth mode of the waveguide with propagation wave number $n_{\mathrm{eff},m}k$. The $\Delta\alpha_{m,\mathrm{RF}}$ seen by the photons entering the region at $z = 0$ is

$$\Delta\alpha_{m,\mathrm{eff}} = \Delta\alpha^{\mathrm{o}}_{m,\mathrm{eff}}\left(F^{\mathrm{o}}_{\mathrm{RF}}\right)e^{-\frac{\alpha_{\mathrm{RF}}}{2}z}\cos[\omega_m t - \delta nkz]. \tag{6.76}$$

The intensity I of these photons after propagating from $z = 0$ to $z = L$ is

$$\frac{I}{I_{\mathrm{o}}} = T(V) = \eta_{\mathrm{ins}}e^{-\Delta\alpha_{\mathrm{bias}}(V_{\mathrm{bias}})L}e^{-\int\limits_0^L \Delta\alpha_{m,\mathrm{eff}}\mathrm{d}z}$$

$$= \eta_{\mathrm{ins}}e^{-\Delta\alpha_{\mathrm{bias}}(V_{\mathrm{bias}})}e^{-\Delta\alpha_{\mathrm{RF}}L\cos(\omega_m t - \xi)}, \tag{6.77}$$

$$\Delta\alpha_{\mathrm{RF}}L\cos(\omega_m t - \xi) = \Delta\alpha^{\mathrm{o}}_{m,\mathrm{eff}}\left(F^{\mathrm{o}}_{\mathrm{RF}}\right)LA\cos(\omega_m t - \xi)$$

$$= \Delta\alpha^{\mathrm{o}}_{m,\mathrm{eff}}\left(F^{\mathrm{o}}_{\mathrm{RF}}\right)\int\limits_0^L e^{-\frac{\alpha_{\mathrm{RF}}}{2}z}\cos(\omega_m t - \delta nkz)\mathrm{d}z. \tag{6.78}$$

Note that I_{o} is the intensity of the guided wave at $z = 0$ and η_{ins} is the insertion efficiency which includes the residual propagation loss of the waveguide. Here we have designated

$$AL\cos(\omega_m t - \xi) = \int\limits_0^L e^{-\frac{\alpha_{\mathrm{RF}}}{2}z}\cos(\omega_m t - \delta nkz)\mathrm{d}z. \tag{6.79}$$

When we carry out the integration, we obtain

$$A^2 = \frac{1/L^2}{\left[(\alpha_{\mathrm{RF}}/2)^2 + (\delta nk)^2\right]}\left\{\left[1 + e^{-\alpha_{\mathrm{RF}}L}\right] - 2\cos(\delta nkL)e^{-\frac{\alpha_{\mathrm{RF}}L}{2}}\right\}, \tag{6.80}$$

$$\tan\xi = \frac{\left(-\dfrac{\alpha_{\mathrm{RF}}}{2}\sin(\delta nkL) - (\delta nk)\cos(\delta nkL)\right)e^{-\frac{\alpha_{\mathrm{RF}}L}{2}} + \delta nk}{\left(-\dfrac{\alpha_{\mathrm{RF}}}{2}\sin(\delta nkL) + (\delta nk)\cos(\delta nkL)\right)e^{-\frac{\alpha_{\mathrm{RF}}L}{2}} + \dfrac{\alpha_{\mathrm{RF}}}{2}}. \tag{6.81}$$

As α_{RF} and δn approach 0, $A \approx 1\ and\ \xi \approx 0$. Therefore, the normalized transmission function $T(V)$ of an ideal traveling wave EA modulator is again identical to the $T(V)$ at low frequencies. All design considerations as well as the pros and cons of lumped element EA modulation discussed in Section 6.2.1.4 are directly applicable to traveling wave EA modulators. However, there is no electrical bandwidth limitation caused by RC_m. In reality, α_{RF} is a function of ω_m. Even for an electrode transmission line well matched to the RF source and the load, α_{RF} and δn limit the $V_{\pi,\mathrm{eq}}$ that can be lowered by lengthening L. In comparison with MZ and directional coupler modulators, the length L

of a typical EA modulator is 1 mm or less while the length of MZ modulators is in centimeters. Therefore the effect of a moderate δn is not as severe. Since EA modulators are made on III-V semiconductor materials that have optical refractive indices in the range 3.2–3.5, a moderate δn is easy to achieve. Irmscher *et al.* have demonstrated an InP-InGaAsP Frantz–Keldysh traveling wave modulator at the 1.55 μm wavelength for digital applications [30]. The impedance of the 12 μm wide electrode is only 11 Ω. It is terminated by a 12 Ω load impedance. For devices 250, 450 and 950 μm long and contrast ratio of 20 dB, the driving voltages are 2, 1.3 and 0.7 V respectively. The devices showed a measured bandwidth of 45–17 GHz. However, the 250 μm long device is predicted to have a 67 GHz bandwidth.

Note that, ideally, electrodes should have a characteristic impedance that is matched to the 50 Ω source impedance, a small δn, a low α_{RF} and a field distribution that optimizes $\Delta\alpha_{m,eff}$ for a given applied voltage. In practice, the requirement to meet all these demands is difficult. How to design and fabricate the traveling wave electrode is an important issue for traveling wave EA modulators. However, because of the short device length, the demand on small δn is not as tough for EA modulators as for long MZ electro-optic modulators.

6.2.2.3   Design of traveling wave electrodes

In Section 4.2.2 we have shown that the transmission line's voltage and current follow a set of one-dimensional equations,[29]

$$\frac{dV}{dz} = -Z_L I \quad \text{and} \quad \frac{dI}{dz} = -Y_c V, \tag{6.82}$$

where

$$Z_L = R_c + j\omega_m L \quad \text{and} \quad Y_c = j\omega_m C + \frac{1}{R_j}. \tag{6.83}$$

The solutions are

$$V(z,t) = [V^f e^{-\gamma z} + V^b e^{+\gamma z}] e^{j\omega_m t} \quad \text{and} \quad I(z,t) = [I^f e^{-\gamma z} + I^b e^{+\gamma z}] e^{j\omega_m t}. \tag{6.84}$$

The first term represents a wave propagating in the positive z direction and the second term a wave propagating in the negative z direction. Their temporal components are allowed to contain separate phase offsets for each of the propagating directions. The equivalent circuit relates I and V through

$$I(z,t) = \frac{1}{Z_o}[V^f e^{-\gamma z} - V^b e^{-\gamma z}] e^{j\omega_m t}, \tag{6.85}$$

where Z_o is the characteristic impedance of the line which is related to the circuit parameters by

$$Z_o = \sqrt{\frac{R_c + j\omega_m L}{1/R_j + j\omega_m C}} = \frac{\gamma}{1/R_j + j\omega_m C}. \tag{6.86}$$

The propagation constant γ is also related to the circuit parameters and given by

$$\gamma = \sqrt{(R_c + j\omega_m L)\left(\frac{1}{R_j} + j\omega_m C\right)} = \frac{\alpha_{RF}}{2} + j\beta_m, \quad (6.87)$$

where R_c represents the conduction loss of the electrodes and R_j represents the parallel leakage resistance between the electrodes. The phase velocity of the microwaves v_p and the group velocity of the microwave pulses v_g propagating in the z direction are

$$v_p = \frac{\omega_m}{\beta_m}, \quad (6.88)$$

$$v_g = \left(\frac{d\omega_m}{d\beta_m}\right). \quad (6.89)$$

Note that α_{RF} is the attenuation coefficient of the intensity of the microwave, and β_m is related to the effective index of the microwave on the transmission line by

$$n_m = \beta_m/k = c/v_p. \quad (6.90)$$

The design of the transmission line is focused on matching the n_m with the n_{eff} of the optical guided-wave, maximizing the electric field in the electro-optic or electro-absorption medium, minimizing α_{RF} and obtaining a Z_o that has an impedance as close to 50 Ω in resistance as possible. Commonly used electrodes are usually in the form of micro-strip MCS, coplanar waveguide CPW and coplanar strip CPS transmission line. For electro-optic modulators on insulators, CPW and CPS are used commonly, while MCS is used commonly for semiconductor modulators. In LiNbO$_3$ modulators, the anisotropic dielectric constants of the material at microwave frequencies are much larger than that at the optical frequencies. In addition to the other design requirements, close matching of n_m with n_{eff} over a long length presents a challenge. In polymer modulators the difference of dielectric constant at microwave and optical frequencies is much smaller. This reduces the problems encountered in transmission line design. Since electro-optic modulators are fabricated on insulators, $1/R_j$ is zero. In semiconductor modulators, the matching of n_m with n_{eff} is easy because the dielectric constants at microwave and optical frequencies are similar. However, the thin intrinsic layer in the reverse biased p–i–n structure yields a large capacitance per unit length C unless the width of the electrode is very small. Impedance matching and microwave attenuation become the primary considerations.

In all modulators the conductor loss R_c is a major concern that affects the α_{RF}, especially at millimeter wave frequencies.[30] Crude approximations for the series resistance R_c may be obtained by calculating the resistance per unit length of the conductors, using the Wheeler incremental inductance rule [see 31–33]. In general

$$\alpha_{RF} = \alpha_o \sqrt{f}, \quad (6.91)$$

and α_o will vary as a function of electrode thickness, width and gap. The calculated α_o as a function of a gold electrode 3 μm thick for CPW and CPS transmission lines with gap g and width w has been given by Chung [34]. The Wheeler approximation is valid only when the conductor thickness is much greater than its skin depth, given by $\sqrt{2/\omega_m\mu\sigma}$.

If we neglect R_c or α_{RF} for the time being, we obtain

$$Z_o = \sqrt{\frac{L}{C}} = \frac{\beta_m}{\omega_m C}. \tag{6.92}$$

Note that guided-wave modulators and switches usually have a large C. Therefore, in order to get a reasonably large Z_o, the electrodes need to have a large L.

The fundamental mode of the electrode transmission line is a TEM mode. The magnetic field experiences a uniform permeability throughout the entire structure, no matter what the dielectric constants for the electric field. If we replace the dielectric medium by air, the inductance L per unit length does not change, while the capacitance per unit length is changed to C_a. The propagation wave number for the TEM mode of all parallel electrodes in air is just k, $k = \omega_m/c = \sqrt{LC_a}$. Thus

$$\beta_m = k\sqrt{\frac{C}{C_a}}, \tag{6.93}$$

and

$$Z_o = \frac{1}{c\sqrt{CC_a}}, \tag{6.94}$$

$$n_m = \sqrt{\frac{C}{C_a}}. \tag{6.95}$$

In order to reduce n_m to match n_{eff}, C/C_a needs to be reduced.

The CPW and CPS transmission lines in LiNbO$_3$ modulators represent a good example for illustrating the design of electrodes. Applying a low index buffer dielectric layer such as SiO$_2$ under the electrode decreases the C as the electric field moves out from the high index LiNbO$_3$. Increasing the thickness of the electrode increases C_a.[31] Note that C and C_a also depend strongly on the width to gap ratio. Therefore, traveling wave LiNbO$_3$ modulators have very thick gold electrodes that have appropriate width and gap. However, the increase of the SiO$_2$ buffer and gold electrode thickness will decrease $V_\pi L_p$ because they decrease the electric field in the electro-optic medium. Howerton and Burns have presented a good review of traveling wave LiNbO$_3$ modulators [35]. Figure 5.10 in this reference shows the effect of varying the gold electrode thickness t_g and the SiO$_2$ buffer thickness t_b on the $\Delta\omega_m L_p$ product, the Z_o and $V_\pi L_p$ for a phase or MZ modulator with CPS electrode on z-cut LiNbO$_3$.[32]

Notes

1. In comparison, planar guided waves are usually millimeters wide. Electro-optical effects over a wide guided-wave beam can only be obtained with electrodes that require large applied voltage, or that have large electrical capacitance. Devices that require large voltage or that have large capacitance can be driven effectively only at low electrical frequencies.

2. Because of the small transverse dimension of the channel waveguide mode, it is possible, but inefficient, to excite the channel waveguide modes by prism or grating as can be done for planar waveguides discussed in Sections 5.1.2.2 to 5.1.2.4. End excitation is the dominant method of coupling the waveguides to the fibers.

3. From the super mode point of view, the mode in the input waveguide is symmetrical. It excites only the symmetric mode at the output of a symmetrical Y-branch, independent of the waveguide configuration. Thus the 3 dB power splitting is tolerant with respect to variations in waveguide and tapering configurations.

4. If there are M inputs, then the Star coupler becomes an $M \times N$ coupler. The phase, amplitude, and frequency of the diffracted fields at the entrance of the output waveguides are determined from the incident field of the input waveguides. There will be interference effects produced from different input signals.

5. Note that $\Delta\phi_Y$ could also be tuned electro-optically by applying a DC electric field to the waveguide.

6. From the point of view of super modes, a two-mode interference coupler is identical to a directional coupler with zero gap between the waveguides.

7. Electro-optical tuning of the relative phase may be used.

8. See Section 5.2.1 for a discussion on various orders of a diffraction grating.

9. For a given Y intersection, the relative phase of the backward propagating modes reaching the input is the sum of the relative phase in the transition region (i.e. the coupled region of the Y intersection) and in the two-mode waveguide. There is no relative phase shift in the uncoupled region.

10. The linewidth $\Delta\omega$ is defined in many optics books as the full linewidth when the amplitude of the transmitted wave drops to half of its maximum. The linewidth in Eq. (6.17) is a half linewidth for the transmitted power to drop to half of its maximum. This difference accounts for the factor $\sqrt{2}$.

11. It is clear from Sections 2.2.2 and 2.2.4 that ϕ_t is determined primarily by $\exp(-jn_{\text{eff}}kL_c)$ at the resonance frequency in the coupling region.

12. See Section 5.5 for a discussion of the acousto-optical scanner, deflector and spectrum analyzer.

13. In the lumped element approximation, the time variation of Δn_{eff} is the same as V at the input, independent of z.

14. The fringe electric field is neglected in this approximation.

15. In order to simplify our discussion, phase modulation is applied only on one arm of the MZ modulator in Fig. 6.4. In practice, more efficient modulation is obtained by obtaining $+\Delta\phi$ in one arm and $-\Delta\phi$ in the other arm, called push-pull operation. The electrode configuration will be different.

16. Applying electrodes on both arms may also double the electrical capacitance of the electrodes to be driven by the RF source at low frequencies. At high frequencies, the transmission lines of the electrodes on both arms may be driven in parallel.

17. For a given RF signal current generated from the detector, the RF power transmitted to the load and the bandwidth will also depend on the RC time constant of the detector circuit. However, since the detector capacitance is much smaller than C_m, the bandwidth of the analog link is determined usually by the modulator C_m.

18. Since the symbol α is used uniformly to designate chirping in the literature, we will also adopt it here. The reader should not confuse it with the absorption coefficient α.

19. Note that $V_\pi L_p$ is usually used to compare different designs of MZ modulator.

20. The C_m of a MZ modulator can be calculated in the same way as the C_m of the phase modulator discussed in Section 6.2.1.1. Note that the C_m of a balanced MZ modulator in push-pull operation is the total C_m of the electrodes for both arms.

21. Since the intensity is proportional to E^2, the absorption coefficient of the intensity of the guided wave is α_m which is $\alpha_{m,o}$ plus $\Delta\alpha_m$. In the lumped element approximation, $\Delta\alpha_m$ for a modulator with constant transverse configuration is independent of z.

22. For quantum well materials, $\underline{F}$ is oriented in the direction perpendicular to the quantum well layers.
23. Strictly, electro-absorption occurs only in the quantum wells, not in the barriers. Since both the wells and the barriers are much thinner than the wavelength of the radiation, the measured α is usually the averaged α of the absorptive medium, including both the wells and the barrier.
24. The SFDR in dB is usually obtained from a log-log plot of RF and distortion at the output versus RF input. How SFDR will change as a function of the noise level and the output distortion depends on the dominant order of non-linearity. The exponent on the units for SFDR reflects the dependence on the order of the dominant non-linearity. See ref. 20 for a detailed discussion.
25. When $\delta n k L_p / 2 = 1.9$, $\Delta\phi$ is reduced to half of its maximum value at $\omega_m = 0$. Therefore, the bandwidth $\Delta\omega$ will depend on the variation of the $\Delta\phi$ allowed.
26. Typically α_{RF} is proportional to $\sqrt{\omega}$ in a microwave transmission line.
27. See also Eq. (6.79) to (6.81) for an explicit expression for $\int_0^{L_p} e^{-\frac{\alpha_{RF}}{2}z} \cos(\omega t - \delta n k z) dz$.
28. According to Section 3.2, α is a non-linear function of F, and $F = F_b + F_{RF}(x,y,z,t)$. Then F_{RF} can be treated as a small signal perturbation of F_b. So $\alpha(F)$ can be expressed by a Taylor series expansion in terms of $F_{RF}(x,y,z,t)$ around F_b. The $\Delta\alpha_{RF}(x,y,z,t)$ in Eq. (6.74) is the dominant first order term in the Taylor series expansion. It is an approximation. There are other higher order terms corresponding to $\cos^n(\omega t - \beta_{RF}z)$ for $n > 1$. They are much smaller.
29. In some transmission lines, there is dielectric loss. In semiconductor transmission lines that include a p–i–n junction there is also contact resistance R_s. These losses have not been included in the transmission line equations in Eq. (6.83).
30. When the thickness of the substrate of the optical waveguide can support microwave slab modes, coplanar transmission lines could also lose microwave energy to the substrate when the effective indices of the slab modes are phase matched to the n_m of the transmission line. For this reason, the $LiNbO_3$ substrate is thinned to reduce α_{RF}.
31. A thicker electrode also has a larger inductance L.
32. The $\Delta\omega_m L_p$ product for a phase modulator is given in Eq. (6.71) for $\Delta\phi$ reduced to half of its maximum value at $\omega_m = 0$. It may vary, depending on the definition of bandwidth. Results given in this figure are applicable to both phase and MZ modulators.

References

1. C. Dragone, Efficient MxN coupler using Fourier optics. *J. Lightwave Tech.*, **7** (1989) 479.
2. L. B. Soldano and E. C. M. Pennings, Optical multi-mode interference devices based on self-imaging: Principles and applications. *J. Lightwave Tech.*, **13** (1995) 615.
3. M. K. Smit and C. van Dam, Phasar based WDM-devices, principles, design, and applications. *IEEE J. Select. Topics Quant. Elect.*, **2** (1996) 236.
4. J. X. Chen, T. Kawanishi, K. Higuma, *et al.*, Tunable lithium niobate waveguide loop. *IEEE Phot. Tech. Lett.*, **16** (2004) 2090.
5. A. Yariv, *Optical Electronics in Modern Communication*, Section 4.1, Oxford University Press (1997).
6. A. Yariv, Universal relations for coupling of optical power between microresonators and dielectric waveguide. *Elect. Lett.*, **36** (2000) 321.
7. T. Kominato, Y. Ohmori, N. Takato, H. Okazaki, and M. Yasu, Ring resonators composed of GeO_2-doped silica waveguide. *J. Lightwave Tech.*, **10** (1992) 1781.
8. Kauhiro Oda, Norio Takato, and Hiroma Toba, A wide FSR waveguide double-ring resonator for optical FDM transmission systems. *J. Lightwave Tech.*, **9** (1991) 728.

9. S. Suzuki, K. Oda, and Y. Hibino, Integrated-optic double-ring resonators with a wide free spectra range of 100 GHz. *J. Lightwave Tech.*, **13** (1995) 1766.

10. H. G. Unger, *Planar Optical Waveguides and Fibers,* Section 2.8, Oxford University Press (1977).

11. G. Lens, B. J. Eggleton, C. K. Madsen, and R. F. Slusher, Optical delay lines based on optical filters. *IEEE J. Quant. Elect.*, **37** (2001) 525.

12. L. Zhuang, C. G. H. Roeloffzen, R. G. Heideman, *et al.*, Single-chip ring resonator-based 1 × 8 optical beam forming network in CMOS-compatible waveguide technology. *IEEE Phot. Tech. Lett.*, **19** (2007) 1130.

13. D. M. Pozar, *Microwave Engineering*, John Wiley and Sons (2005).

14. H. Chung, W. S. C. Chang, and E. L. Adler, Modeling and optimization of traveling-wave electrode in LiNbO$_3$ electro-optic modulators. *IEEE J. Quant. Elect.*, **27** (1991) 608.

15. G. E. Betts, LiNbO$_3$ external modulators and high performance analog links, Chapter 4 in *RF Photonic Technology in Optical Fiber Links*, ed. W. S. C. Chang, Cambridge University Press (2002).

16. F. Koyama and K. Iga, Frequency chirping in external modulators. *J. Lightwave Tech.*, **6** (1988) 87.

17. A. H. Gnauck, S. K. Korotky, J. J. Veselk, *et al.*, Dispersion penalty reduction using an optical modulator with adustable chirp. *IEEE Phot. Tech. Lett.*, **3** (1991) 916.

18. K. K. Loi, Multiple-quantum-well waveguide modulators at 1.3 μm wavelength. Ph.D. thesis, University of California San Diego (1998).

19. K. K. Loi, J. H. Hodiak, X. B. Mei, *et al.*, Low-loss 1.3 μm MQW electro-absorption modulators for high-linearity analog optical links. *IEEE Phot. Tech. Lett.*, **10** (1998) 1572.

20. Y. Zhuang, Peripheral coupled waveguide multiple quantum well electro-absorption modulator for high efficiency, high spurious free dynamic range and high frequency fiber optical link. Ph.D. thesis, University of California San Diego (2005).

21. Y. Zhuang, W. S. C. Chang, and P. K. L. Yu, Peripheral-coupled-waveguide MQW electro-absorption modulator for near transparency and high spurious free dynamic range RF fiber-optic link. *IEEE Phot. Tech. Lett.*, **16** (2004) 2033.

22. G. E. Betts, X. Xie, I. Shubin, W. S. C. Chang, and P. K. L. Yu, Gain limit in analog links using electro-absorption modulators. *IEEE Phot. Tech. Lett.*, **18** (2006) 2065.

23. J. A. J. Fells, I. H. White, M. A. Gibbon, *et al.*, Controlling the chirp in electro-absorption modulators under digital modulation. *Elect. Lett.*, **30** (1994) 2066.

24. F. Devaux, Y. Sorel, and J. F. Kerdiles, Simple measurement of fiber dispersion and of chirp parameter of intensity modulated light emitter. *J. Lightwave Tech.*, **11** (1993) 1937.

25. T. Ido, S. Tanaka, M. Suzuki, *et al.*, Ultra-high-speed multiple-quantum-well optical modulators with integrated waveguides. *J. Lightwave Tech.*, **14** (1996) 2026.

26. H. Kawanishi, Y. Yamauchi, N. Mineo, *et al.*, EAM-integrated DFB laser modules with more than 40-GHz bandwidth. *IEEE Phot. Tech. Lett.*, **13** (2001) 954.

27. M. W. Howerton, R. P. Moeller, A. S. Greenblatt, and R. Krahenbuhl, Fully packaged, broadband LiNbO$_3$ modulator with low drive voltage. *IEEE Phot. Tech. Lett.*, **12** (2000) 792.

28. K. Noguchi, G. Motomi, and H. Miyazawa, Millimeter-wave Ti-LiNbO$_3$ optical modulators. *J. Lightwave Tech.*, **16** (1998) 615.

29. S. K. Korotky, G. Eisenstein, R. S. Tucker, J. J. Veselka, and G. Raybon, Optical intensity modulation to 40 GHz using a waveguide electro-optic switch. *Appl. Phys. Lett.*, **50** (1987) 1631.

30. S. Irmscher, R. Lewen, and U. Eriksson, InP-InGaAsP high-speed traveling-wave electro-absorption modulator with integrated termination resistors. *IEEE Phot. Tech. Lett.*, **14** (2002) 923.

31. H. A. Wheeler, Formulas for the skin effect. *Proc. IRE*, **30** (1942) 412.
32. R. A. Purcel, D. J. Masse, and C. P. Hartwig, Losses in microstrip. *IEEE Trans. Microwave Theory and Tech.*, **MTT-16** (1968) 342.
33. E. J. Denlinger, Losses of microstrip lines. *IEEE Trans. Microwave Theory and Tech.*, **MTT-28** (1980) 513.
34. H. Chung, Optimization of microwave frequency traveling-wave LiNbO$_3$ integrated-optic modulators. Ph.D. thesis, University of California San Diego (1990).
35. M. M. Howerton and W. K. Burns, Broadband traveling wave modulators in LiNbO$_3$, Chapter 5 in *RF Photonic Technology in Optical Fiber Links*, ed. W. S. C. Chang, Cambridge University Press (2002).

Index